"We find them smaller and fainter, in constantly increasing numbers, and we know that we are reaching into space, farther and farther, until, with the faintest nebulae that can be detected with the greatest telescopes, we arrive at the frontier of the known universe."

— Edwin Powell Hubble

We acknowledge the financial support of the Government of Canada through the Book Publishing Industry Development Program for our publishing activities.

Published by Apogee Books, Box 62034, Burlington, Ontario, Canada, L7R 4K2, http://www.apogeebooks.com
Tel: (905) 637-5737
Printed and bound in Canada
Hubble Space Telescope Pocket Space Guide by Steve Whitfield
Apogee Books Pocket Space Guide #7

ISBN-10: 1-894959-38-8
ISBN-13: 978-1-894959-38-4

Hubble Space Telescope

Pocket Space Guide

by Steve Whitfield

Contents

Introduction

Historically, the space activities that have most captured the public's attention have revolved around a *mission* – a set of activities and milestones that were entirely preplanned, in detail, and were then executed, hopefully, entirely as rehearsed, much in the manner of a stage production. The situation with the Hubble Space Telescope is quite different. Hubble began with a broad set of general goals, and the results of its activities were used to determine what the details of its subsequent activities would be.

Another difference is the time frame. The significant events of a typical manned mission have collectively required at most a few days to execute, and for an unmanned mission perhaps a few weeks. The Hubble program was planned to be 15 years in duration, was then extended to 20 years, and current discussions may extend that even further. Hubble has been up there ticking away for 16 years so far.

Telescope History

Galileo Galilei (1564 - 1642) was the first person to point a telescope skyward. Although his telescopic images were fuzzy and small, Galileo was able to make out mountains and craters on the Moon, four moons orbiting Jupiter, and a ribbon of diffuse light across the sky which his successors would identify as our Milky Way galaxy.

Galileo Galilei

Over time, larger and more complex telescopes were produced; astronomers discovered many faint stars and began to calculate stellar distances. In the 19th century, a new instrument called a spectroscope was used with telescopes to gather information about the chemical composition and motions of celestial objects.

In the year 1929, an astronomer named Edwin Hubble changed our view of the universe. Using his era's largest telescope, he discovers that beyond our Milky Way are billions of other galaxies, all moving away from us. 20th century astronomers developed bigger and better telescopes, continually extending man's view into the universe.

In 1968, the Orbiting Astronomical Observatory, precursor to the Hubble Space Telescope, entered orbit. Its observations were a stunning success. Throughout the early 1970s NASA and the astronomical community begin planning and building support for a large telescope in Earth orbit, where it would be free of the distortions caused by light passing through Earth's atmosphere. The result was the Hubble Space Telescope, launched in 1990.

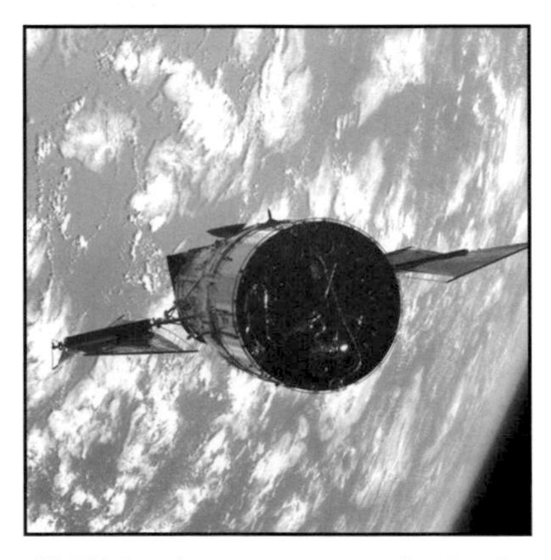

Hubble's solar arrays seen as the shuttle approaches prior to capture.

Hubble completes one orbit around the Earth every 97 minutes, orbiting about 600 km (380 miles) above the surface. The furthest objects that Hubble has seen are galaxies well over 12 billion light-years away. On a dark, clear night, Hubble can be seen with the naked eye as it passes directly over head.

Who Was Edwin Hubble?

Edwin Powell Hubble (1889-1953) was an American astronomer best known for discovering in 1924 that there are galaxies other than our own Milky Way. In 1927 he determined that all of these galaxies were receding from us at rates that increased with their distance from our galaxy. This was the first evidence that our universe is not static, as had previously been thought, but was in fact expanding.

Edwin Powell Hubble.

Ten years earlier Albert Einstein's General Theory of Relativity had indicated that the universe must be either expanding or contracting. Einstein wasn't happy with this idea and so he invented a "cosmological constant" which eliminated the expanding or contracting requirement from relativity theory. When Hubble's discovery validated the expanding universe concept, Einstein reflected on having added the cosmological constant to his equations and referred to it as "the biggest blunder of my life."

Hubble also developed a classification system for galaxies, in which they were grouped according to their content, distance, shape, size and brightness.

Edwin Hubble is often credited with having discovered the red shift of the galaxies. In actual fact, the red shift was known and understood prior to Hubble's time. What he did accomplish, along with Milton Humason, was to determine that the red shifting of each galaxy is proportional to its distance from us.

Their empirical Redshift Distance Law of galaxies is now generally referred to simply as Hubble's law. This law shows that the farther two galaxies are from each other, the faster they are moving away from each other. It was this discovery that led to Big Bang theory.

In addition to the Hubble Space Telescope, Edwin Hubble has had and asteroid (2069 Hubble) and a crater on the Moon named for him.

Chronology of the Hubble Space Telescope

1946: Lyman Spitzer is the first to push for a space-based telescope.
1969: The National Academy of Sciences urges NASA to develop an orbital telescope.
1971: NASA creates the Large Space Telescope Science Steering Group to begin feasibility studies.
1975: ESA become involved with the program.
1975: The diameter of the telescope is reduced from 3 meters to 2.4 meters.
1977: Initial designs are refined; the program receives funding from Congress; and NASA awards contracts.
1981: The telescope's mirror is completed. It will later prove to have problems which cause blurred images.
1981: NASA forms the Space Telescope Science Institute to manage the telescope's science program.
1983: The spacecraft is officially named after astronomer Edwin P. Hubble.
1984: Assembly of the optical system completed.
1985: Final assembly of the spacecraft is completed.
1986: Launch is scheduled for autumn, but later delayed by the *Challenger* disaster.

1990: Hubble is launched aboard *Discovery* (STS-31) on April 24. In June, scientists uncover spherical aberration in the mirror that causes slightly blurred images.

1993: Servicing Mission 1 in December (STS-61) improves Hubble's image quality by installing corrective optics.

1994: Hubble observes pieces of the Comet Shoemaker-Levy 9 plunge into the atmosphere of Jupiter.

1997: Servicing Mission 2 in February (STS-82), replaces several of Hubble's parts.

1998: Hubble research indicates that expansion of the universe is accelerating, not slowing down.

1999: Servicing Mission 3A in December (STS-103) replaces all six of Hubble's gyroscopes.

2002: Servicing Mission 3B in March (STS-109) replaced the solar arrays for a third time and installs the Advanced Camera for Surveys.

2004: The loss of *Columbia* in early 2003 makes another servicing mission by the shuttle unlikely. A robotic servicing mission to repair Hubble is deemed too costly.

2005: NASA Administrator Michael Griffin revisits the decision to cancel a servicing mission and commissions additional studies at Goddard Space Flight Center.

Hubble at a Glance

Dates			
Launched:	April 24, 1990		
Deployed:	April 25, 1990		
First Image:	May 20, 1990		
Servicing Mission 1:	December 1993		
Servicing Mission 2:	February 1997		
Servicing Mission 3A:	December 1999		
Servicing Mission 3B:	February 2002		
Mission Duration:	Up to 20 years		
Orbit			
Orbital Speed:	4.85 miles/sec	17,500 mph	28,100 kph
Orbital Altitude:	380 miles	600 km	
Orbital Inclination:	28.5 degrees to the equator		
Orbital Duration:	97 minutes		
Mechanical			
Length:	43.5 feet	13.2 m	
Diameter:	14 feet	4.2 m	
Weight:	24,500 lb	11,110 kg	
Primary Mirror Diameter:	94.5 inches	2.4 m	
Primary Mirror Weight:	1,825 pounds	828 kg	
Secondary Mirror Diameter:	12 inches	0.3 m	
Secondary Mirror Weight:	27.4 pounds	12.3 kg	
Electrical			
Sensitivity:	Ultraviolet (115 nm) through infrared (2500 nm).		
Data Transmission:	About 120 gigabytes per week.		
Energy source:	The Sun via two 25-foot solar panels		
Power usage:	2,800 watts		
Batteries:	6 nickel-hydrogen		

What Does the Hubble Space Telescope Do?

Hubble's "Job Description" has been listed as:

Explore the solar system.
Measure the age and size of the universe.
Search for our cosmic roots.
Chart the evolution of the universe.
Unlock the mysteries of galaxies, stars, planets, and life itself.

From its unique vantage point 600 kilometers (380 miles) above the surface of the Earth, the Hubble Space Telescope (HST) looks out into space with a resolution from 120 nanometers (near-ultraviolet) to 2500 nanometers (near-infrared). Earth's atmosphere absorbs a great deal of ultraviolet and infrared radiation, and distorts visible light images as well. HST is able to capture images and spectra from distant stars which would be difficult or impossible to obtain from the ground.

Conceptually, the Hubble Space Telescope performs the same function as did Galileo's telescope in 1609 – light from extraterrestrial objects is magnified and recorded. Where Galileo used his eyes, pen and paper to record his sightings, Hubble uses electronic cameras and data transmission.

Galileo's observations were limited to light visible to the human eye, but Hubble can "see" and record wavelengths from ultraviolet (115 nanometers wavelength) through infrared (2500 nanometers wavelength). This allows Hubble's images to convey information about temperature, density, electronic structure, and much more from objects across the entire visible universe.

Hubble's optics, science instruments and spacecraft systems work together to capture light and other radiation from the universe, convert it into digital data, and transmit it back to

Earth. Brief descriptions of Hubble's optics and science instruments are given below. Detailed discussions can be found in *Scientific Instruments* section of this book.

Looking into Time

The light that Hubble sees may have taken minutes to travel from its source, or billions of years, depending on the distance it has traveled.

Planets: minutes to hours – Hubble can view events in our solar system on the same day that they happen. Light from Mars takes just a few minutes to reach Hubble. Events on Pluto are about five hours old when Hubble sees them.

Stars: years to thousands of years – Our galaxy is about 100,000 light-years across. The region that Hubble can see (not blocked by dust) extends from a few light-years to a few thousand light-years away, so what Hubble sees today happened a few years to a few thousand years in the past.

Galaxies: millions to billions of years – Light from galaxies beyond the Milky Way takes millions to billions of years to reach Hubble. We see these other galaxies as they were in the far distant past.

Universe: billions of years – Light from the farthest regions of the universe takes billions of years to reach Hubble. The most distant events that Hubble sees occurred when the universe was only a small fraction of its present age.

Optics

The Optical Telescope Assembly (OTA) has four main components:

The *primary mirror* captures light from objects in space and focuses it toward the secondary mirror.
The *secondary mirror* redirects the light coming from the primary mirror onward to the focal plane.

Starlight is focused onto the *focal plane*, where it is picked up by the science instruments.

Corrective optics are added at each science instrument to compensate for an imperfection in the primary mirror.

Science Instruments

Hubble's five science instruments work together and individually to produce images encoded as digital data. Each instrument shows the universe in a unique way. These instruments are as follows:

The Advanced Camera for Surveys (ACS) observes weather on other solar system planets, conducts new surveys of the universe, and studies the nature and distribution of galaxies.

The Wide Field and Planetary Camera (WFPC2) is Hubble's main camera. It is used to observe just about everything.

The Near Infrared Camera and Multi-Object Spectrometer (NICMOS) is Hubble's heat sensor. Its sensitivity to infrared light allows it to observe objects obscured by interstellar gas and dust, so it can peer deep into space.

The Space Telescope Imaging Spectrograph (STIS) acts like a prism to separate captured light into its component colors, which provides data about its temperature, chemical composition, density and motion.

The Fine Guidance Sensors (FGS) are targeting devices that lock onto "guide stars" to keep Hubble pointed in the right direction.

Spacecraft Systems

Hubble obviously needs to be a spacecraft as well as a telescope. Spacecraft support systems are attached to the body of the telescope.

Four *communications antennae* allow astronomers and technicians to communicate with the telescope.

Two *solar arrays* provide electrical power, some of which is stored in onboard batteries so that the telescope can operate while in Earth's shadow.

Several *computers* and *microprocessors* reside in the Hubble body and in each science instrument. Two main Hubble-mounted computers direct the overall operations.

The Hubble *structure and housing* supports and protects the telescope and science instruments. Materials have been chosen to stand up to thermal extremes, and expansion and contraction.

Hubble Program History

In the mid-1960s, NASA and its contractors conducted phased studies into the feasibility of a large space telescope. Some participants argued for an incremental program that would work up to a large-scale observatory, but with the decision to develop the space shuttle, it was decided to take one giant leap to the final telescope / spacecraft. In 1971, approval was given to the Large Space Telescope Science Steering Group to conduct formal feasibility studies.

Once NASA began studies, the next step was to obtain federal funding for what was estimated to be a $400 to $500 million program (the final launch cost was about $1.5 billion). The House Appropriations Subcommittee originally denied funding in 1975. After a price reduction (by using a smaller mirror), much lobbying, and an agreement for participation by ESRO (later to become ESA), Congress granted funding for the Large Space Telescope program in 1977.

Formal design of the telescope began in 1978 with contract awards to Perkin-Elmer Corporation to construct the mirror and optical assembly, and the Lockheed Missiles and Space Company to construct the spacecraft and its support systems. ESA was mainly responsible for the solar array that would power Hubble while in orbit.

NASA's original plan was to launch in 1983, but the program experienced delays. The primary mirror was finished in 1981, but the entire optical assembly was not put together until 1984, and final assembly of the spacecraft did not happen until 1985.

Also in 1985, the new Space Telescope Science Institute at the John Hopkins University was put in charge of Hubble's scientific program. That same year the decision was made to name the telescope for Edwin P. Hubble.

NASA next planned to launch Hubble in October 1986, but the loss of the space shuttle *Challenger* brought major delays to this and many other programs. Shuttle flights resumed in 1988, and Hubble was finally launched aboard *Discovery* (STS-31) on April 24, 1990, and deployed on April 25.

The original program plan was for Hubble to be operational for up to 20 years, but current discussions are for the program to run until at least 2013. The major factor in the actual HST lifetime, of course, is its continued ability to function. There have been four servicing missions to date:

Service Mission 1, December 1993.
Service Mission 2, February 1997.
Service Mission 3A, December 1999.
Service Mission 3B, February 2002.

Service Mission 4 (SM4) has been an on-again, off-again mission for some time. As of June 2006, the mission does not have approval and is contingent on a number of prior mission accomplishments, several risk analysis issues, and, as always, changes in the budget. Meeting these requirements will not guarantee that SM4 will happen. It is still an issue that NASA senior management will decide. NASA's Goddard Space Flight Center continues to work toward a possible SM4 in late 2007 or early 2008.

In June of 1990, it was discovered that Hubble's primary mirror had a spherical aberration (a flaw in its shape) that inhibited its ability to properly focus light, resulting in images that were fuzzy. Later in the year NASA approved COSTAR (Corrective Optics Space Telescope Axial Replacement), a complex packaging of five optical mirror pairs which would rectify the spherical aberration in the primary mirror.

During Servicing Mission 1 (STS-61) the Wide Field Planetary Camera (WFPC1) was replaced by an improved version (WFPC2). This camera system has provided more data and photographs than any other HST instrument.

In addition, SM1 included the installation or replacement of other components including:

Solar Arrays.
Solar Array Drive Electronics (SADE).
Magnetometers.
Coprocessors for the flight computer.
Two Rate Sensor Units.
Two Gyroscope Electronic Control Units.
Goddard High Resolution Spectrograph Redundancy Kit.

Service Mission 2 (STS-82) installed two new instruments, extending Hubble's imaging and spectroscopy wavelength range into the near infrared, and replaced failed or degraded spacecraft components.

NASA decided to split the Third Servicing Mission (SM3) into two parts, SM3A and SM3B, after the third of Hubble's six gyroscopes failed. The failure of a fourth gyroscope on November 13, 1999 placed Hubble in "safe hold" until Service Mission 3A (STS-103) replaced equipment and performed maintenance upgrades. No new scientific instruments were installed. All six of the telescope's aging gyroscopes were replaced. In addition, one of the telescope's three fine guidance sensors was replaced and a new computer was installed.

WFPC2 in the enclosure.

WFPC2 significantly improved ultraviolet performance over WFPC1, the original instrument. In addition to having more advanced detectors and more stringent contamination control, it also incorporated built-in corrective optics.

Service Mission 3B (STS-109) installed the NICMOS Cooling System (NCS) for the Near Infrared Camera and Multi-Object Spectrometer that became dormant in 1999 after depleting its coolant.

In addition, SM3B installed Rigid Solar Arrays (SA3), which are 45% smaller than the first two pairs, but produce 25% more power, and a new Power Control Unit (PCU), to replace the original PCU, which controls and distributes electricity from the solar arrays and batteries to other parts of the telescope.

The summary of scientific instrument installations is given in the following table.

Original Deployment (April 1999)
• Wide Field Planetary Camera 1 (WFPC1). • Goddard High Resolution Spectograph (GHRS). • Faint Object Spectograph (FOS). • Faint Object Camera (FOC). • High Speed Photometer (HSP).
Service Mission 1 (December 1993)
• WFPC1 removed (replaced by WFPC2). • Wide Field Planetary Camera 2 (WFPC2) installed. • HSP removed. • Corrective Optics Space Telescope Axial Replacement (COSTAR) installed to repair primary mirror aberration.
Service Mission 2 (February 1997)
• GHRS removed. • FOS removed. • Space Telescope Imaging Spectograph (STIS) installed. • Near Infrared Camera and Multi-Object Spectrometer (NICMOS) installed.
Service Mission 3A (December 1999)
• NICMOS turned off until NCS installed.
Service Mission 3B (January 2002)
• NICMOS Cooling System (NCS) installed. • NICMOS turned back on. • FOC removed. • Advanced Camera for Surveys (ACS) installed.

(See the Scientific Instruments section of this book for details.)

Hubble Program Status

The Hubble Space Telescope's future became uncertain with the space shuttle *Columbia* tragedy in February 2003. The resulting two-year period of investigations led to then-Administrator Sean O'Keefe's decision that future shuttle missions would go only to the International Space Station, where inspection and repair were possible, and the crew could find safety. The telescope had been scheduled for servicing in 2005, but O'Keefe concluded that another Hubble repair mission would be too risky. In late 2004 the National Academy of Sciences concluded that the risks were acceptable and recommended that another shuttle mission should service Hubble after all. O'Keefe asked NASA's Goddard Space Flight Center to conduct feasibility studies into a robotic servicing mission. Current Administrator Michael Griffin rejected the robotic servicing mission as too costly. Pending successful shuttle operations in the near future, NASA is again considering a mission to Hubble before the shuttle is retired.

Optical Systems

Hubble works on the same principle as the first reflecting telescope built in the 1600s by Isaac Newton. Light enters the telescope and strikes a concave primary mirror, which acts like a lens to focus the light. The bigger the mirror, the better the image.

Hubble's optical telescope assembly consists of two mirrors, support trusses, and the focal plane structure. It is a Ritchey-Chrétien design in which two aspheric (wide angle) mirrors serve to form focused images over the largest possible field of view.

There are three classes of telescope: refractors, which use lenses; reflectors, which use mirrors; and catadioptrics, often called simply cats, which use both mirrors and lenses to focus the incoming light.

Cassegrain telescopes are reflector telescopes that use a combination of two mirrors, called the primary and secondary mirrors, aligned symmetrically about the optical axis. The Ritchey-Chrétien design is a variation of the Cassegrain telescope which uses a hyperboloid mirrors for both primary and secondary, eliminating the corrector plate needed for the catadioptic telescopes. Most modern high-end telescopes, including the Hubble Space Telescope, use the Ritchey-Chrétien design.

Light Path

Incoming light travels down a tube fitted with baffles that keep out stray light. The light is reflected by the concave primary mirror toward the smaller, convex secondary mirror, and then focused on a small area called the focal plane, where it is picked up by the various science instruments.

Figure 1. Light enters Hubble's ***aperture*** and travels down the ***main baffle***. A baffle is a surface which eliminates stray light.

Figure 2. Light is reflected by the ***primary mirror*** which measures about 7 feet (2.4 meters) in diameter. Because of the concave shape, the primary mirror converges the light to the secondary mirror through a ***secondary baffle***.

Figure 3. The ***secondary mirror***, measuring about 1 foot (0.3 m) in diameter, receives the light. It in turn reflects the still-converging light back to the primary mirror through a ***central baffle***.

Figure 4. Light travels through a hole in the primary mirror, to reach the ***focal plane***, where the science instruments examine the light.

Mirror Quality

Hubble's mirrors are very smooth and have precisely shaped reflecting surfaces. Their surfaces do not deviate from a perfect curve by more than 1/800,000ths of an inch. Hubble's

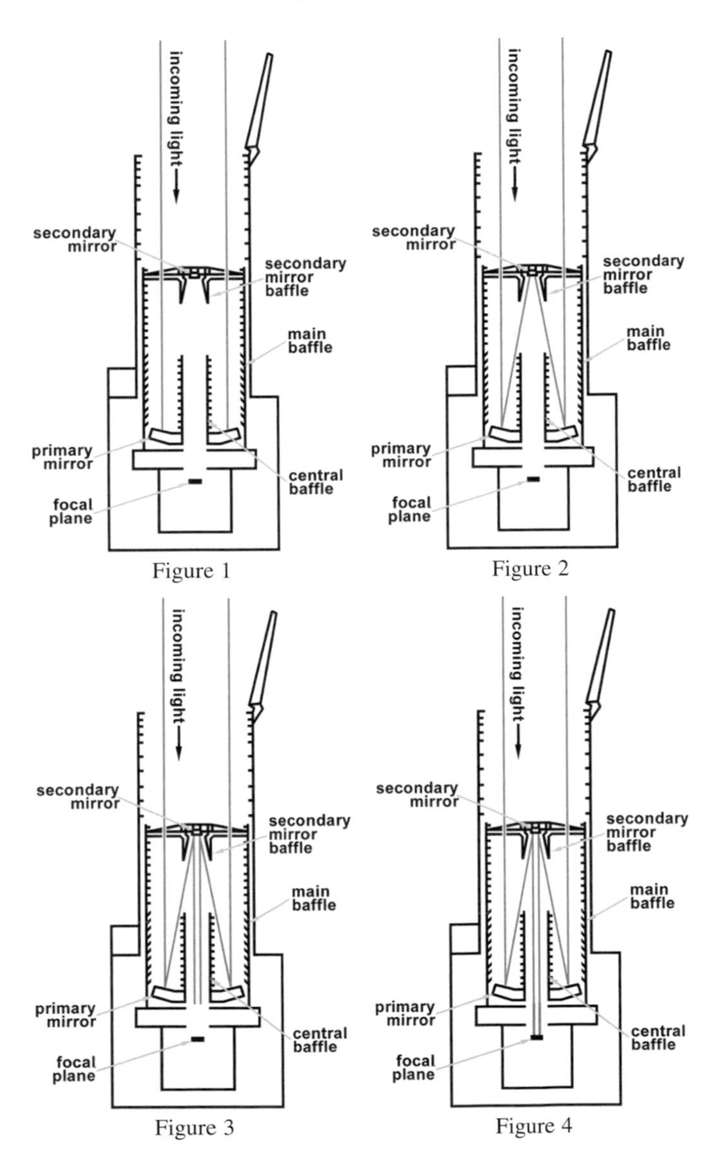

Figure 1

Figure 2

Figure 3

Figure 4

primary mirror required two years to polish off 200 pounds of glass during the grinding of the 94-inch mirror.

Hubble's mirrors are made of ultra-low expansion glass and kept at a nearly constant room temperature (about 70° Fahrenheit) to avoid warping. The reflecting surfaces are coated with a 3/1,000,000th-inch layer of pure aluminum and protected by a 1/1,000,000th-inch layer of magnesium fluoride. The magnesium fluoride makes the mirrors more reflective of ultraviolet light.

The HST optics are among the finest ever made. They are so precise that at wavelengths greater than approximately 300 nanometers, the image quality is limited only by the laws of physics.

Spacecraft Systems

Hubble is not that far away – only about 600 kilometers (380 miles) above Earth's surface. Its nearly circular orbit takes it once around the planet every 97 minutes. The orbit is inclined to the equator at an angle of 28.5 degrees, which means that it never travels more than 28.5° north or south of the equator (Cape Canaveral is at 28.5° north).

The Hubble Space Telescope is, in every sense, a spacecraft. It was designed to be serviced and maintained in orbit, otherwise it wouldn't continue to enjoy its world-class status. The scientific instruments and many other spacecraft parts were made to be replaced by astronauts quickly and easily. These replaceable parts are called Orbital Replacement Units (ORUs) and Orbital Replacement Instruments (ORIs).

HST has eight ORI bays that hold its complement of science instruments. The instruments are classified as "radial" or "axial," depending on position and instrument shape.

Spacecraft Outer Structure

Hubble is designed to operate in space. For this it needs a power supply, communications equipment and a control system. These systems are located around the body of the spacecraft and encompassed by the outer hull.

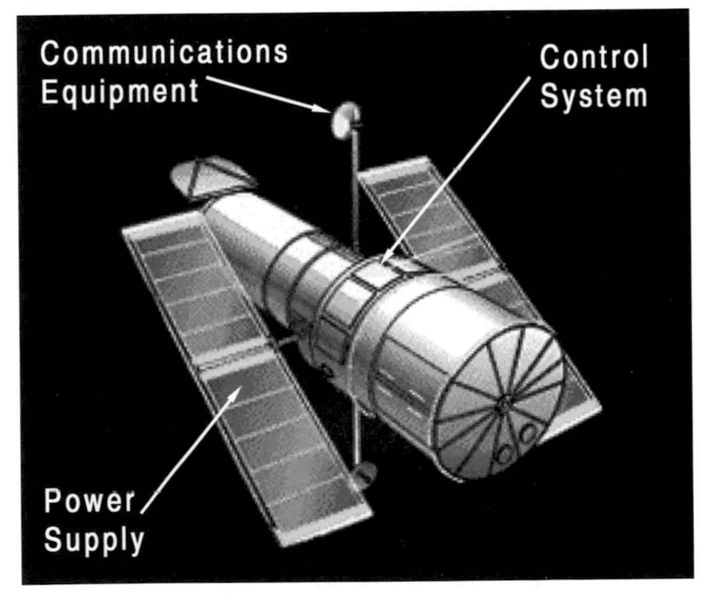

Solar arrays (2) – ESA-supplied Solar panels convert sunlight into 2800 watts of electricity in order to power the telescope. Rechargeable batteries supply back-up power when Hubble is in Earth's shadow (36 minutes out of each 97-minute orbit).

Communications antennae (2) – Steerable antennas send scientific data to orbital communications satellites for relay to ground systems, where the data are stored on solid state recorders.

Computer support systems modules – These modules contain devices and systems needed to operate the Hubble Telescope. They serves as the master control system for communications, navigation, power management, etc.

Electronics boxes – They house much of the electronics, including computer equipment and rechargeable batteries.

Aperture door – The telescope's aperture door protects Hubble's optics in the same way a camera's lens cap shields the lens, closing when Hubble is not in operation.

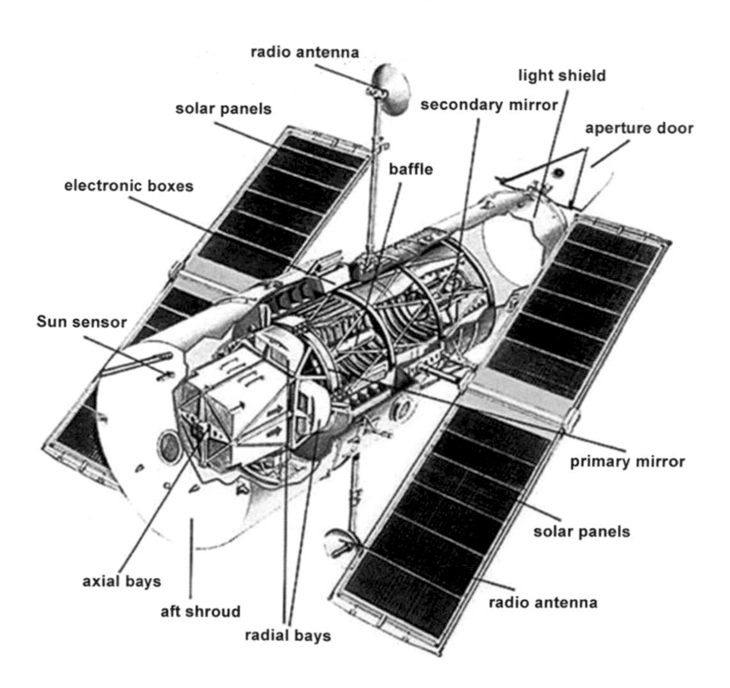

Light shield – Light passes through this shaft before entering the optics system. It blocks surrounding light from entering Hubble.

Hubble's designers had to accommodate the fact that it would be subjected to zero gravity and temperature fluctuations of more than 100 degrees Fahrenheit during each trip around the Earth. Hubble was therefore given a "skin" of multilayered insulation (MLI), to protect it. Beneath the MLI is a lightweight aluminum shell, which provides an external structure to the spacecraft and houses its optical system and science instruments.

The side-mounted solar arrays are designed for replacement by visiting astronauts. They can be folded for shuttle trips to and from Hubble. During maintenance, each fully-charged battery contains enough energy to sustain the telescope in normal science operations mode for 7.5 hours, or five orbits.

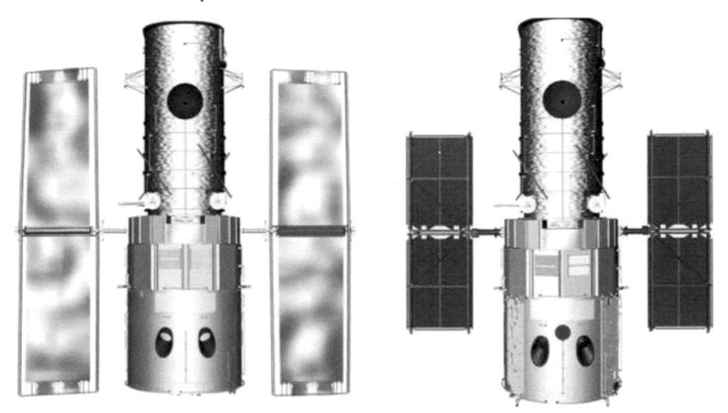

Size comparison of the original solar panels (left) with the smaller but more powerful panels installed later.

Computers and Automation

To run all of the subsystems on board Hubble, computers and microprocessors reside in the Hubble body and in each science instrument. Two main computers direct the show. One main computer talks to the instruments and sends their data to interface units for transmission to the ground, and sends commands and timing information to the instruments. The other main computer handles the gyroscopes, the pointing control subsystem, and other system-wide functions. Special backup computers keep Hubble safe in the event of a problem.

Each instrument itself also houses small computers and microprocessors which direct their activities. These computers direct the rotation of filter wheels, open and close exposure shutters, maintain the temperature of the instruments, collect data, and talk to the main computers.

Pointing Control System

This system aligns the spacecraft to point to and remain locked on a target. Hubble's pointing control system is made up of fine guidance sensors (FGSs), reaction wheels and gyroscopes. There are no rockets on Hubble, because rocket exhaust would contaminate the space near the telescope.

Gyroscopes – Hubble's gyroscopes always face the same direction, like a compass needle. They sense the telescope's angular motion and provide a short-term reference point to help Hubble zero in on its target.

Reaction wheels – Reaction wheels are Hubble's steering system – they spin one way and Hubble will spin the other. Software commands the reaction wheels to spin and accelerating as needed to rotate the telescope to a new target.

Fine Guidance Sensors – Three FGSs are Hubble's targeting devices. They aim the telescope by locking onto "guide stars" and measuring the position of the telescope relative to the target. The sensors provide the precise reference point from which the telescope can begin repositioning.

According to Newton's Third Law of Motion, every action has an equal and opposite reaction. Therefore, as the reaction

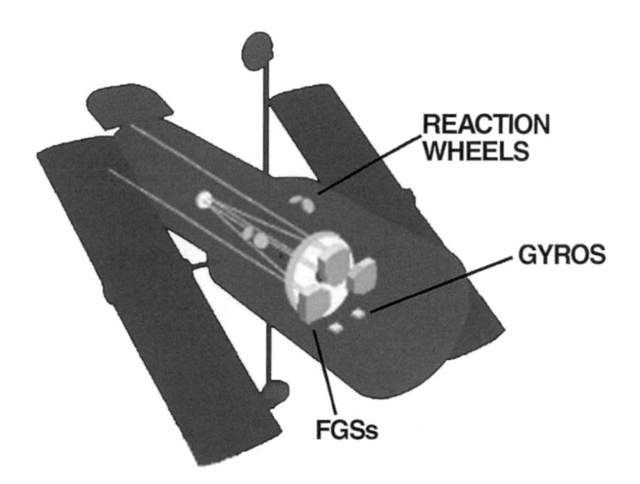

wheels are rotated in one direction, Hubble reacts by rotating in the opposite direction. Since the rotation axes of the four reaction wheels point in different directions, Hubble can point itself toward any location in the sky.

Before Hubble can make an observation, it must find a pair of "guide stars," a star of cataloged exact location to either side of the observation target. To find these directional beacons, mission planners refer to an immense catalog containing the sky "addresses" for 15 million stars. Hubble aims the telescope by locking onto these guide stars and measuring the telescope's position relative to the target.

Two of Hubble's three fine guidance sensors are used for

targeting; the third sensor is available to perform scientific observations.

Resolution

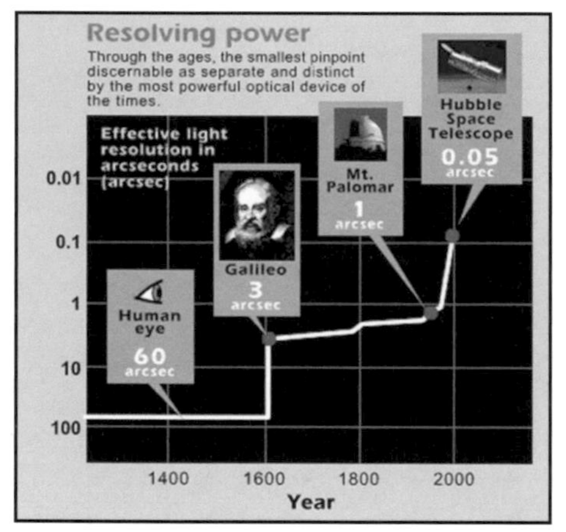

When astronomers talk about how clearly a telescope "sees," they are referring to its resolving power, or resolution – how fine a detail it can see or how close two objects (such as stars) can be together and still be seen as two distinct objects. Astronomers measure the resolving power of a telescope in arcseconds, where:

1 degree = 60 arcminutes = 3600 arcseconds

The best ground-based telescopes can rarely differentiate between two stars that are less than 1 arcsecond apart in the sky. By comparison, Hubble can see detail down to less than 0.1 arcsecond across – more than 10 times more clearly.

Sensors and Actuators

Hubble employs a variety of sensors to detect its own orientation and position. All work in tandem to send the correct information to the actuators to adjust Hubble's position on command.

Fine guidance sensors (3) – These sensors are locked onto two guide stars to keep Hubble in the same relative position of these stars.

Coarse Sun sensors (2) – These sensors measure Hubble's orientation to the Sun, and assist in deciding when to open and close the aperture door.
Magnetic sensing system – These sensors measures Hubble's position relative to Earth's magnetic field.
Rate sensor units (3) – These two rate-sensing gyroscopes measure the attitude rate motion about its sensitive axis.
Fixed head star trackers (3) – A star tracker is an electro-optical detector that locates and tracks a specific star within its field of view.
Reaction wheel actuators (4) – The reaction wheels work by rotating a large flywheel up to 3,000 rpm or braking it to exchange momentum with the spacecraft, which will make Hubble turn.
Magnetic torquers (4) – The torquers are used primarily to manage reaction wheel speed. Reacting against Earth's magnetic field, the torquers reduce the reaction wheel speed, thus managing angular momentum.

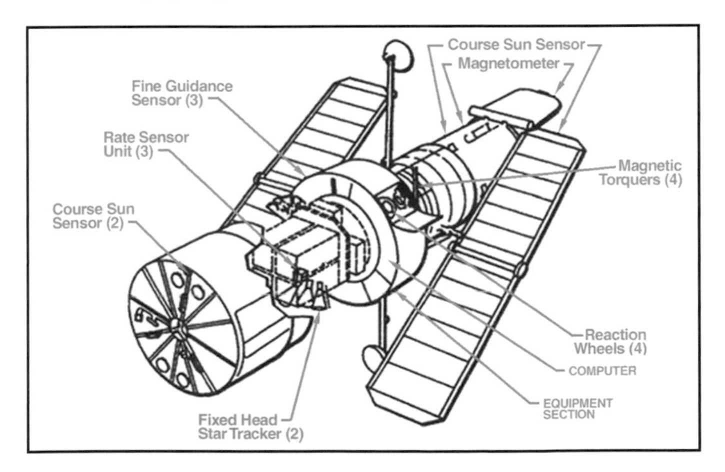

The Tracking and Data Relay Satellite System

Scientists communicate with the Hubble telescope via the Tracking and Data Relay Satellite System (TDRSS) which has satellites located at various locations in the sky. In order for this system to work, at least one of the satellites must be visible from the spacecraft's line of sight. Scientists can interact directly with the telescope during times of satellite visibility, allowing them to make small changes in the spacecraft pointing to fine tune their observations.

Satellite visibility does not affect a planned observation because the commanding is done well in advance. When none of the satellites are visible from the spacecraft, a special data recorder stores the observation. The data are stored and then transmitted during periods of satellite visibility.

To date, ten TDRSS satellites have been launched. The second satellite, TDRS-B, was on board the ill-fated *Challenger* launch (STS-51L) in 1986 and was lost. The other nine satellites are in geosynchronous orbit at various longitudes. The current system utilizes three primary satellites, with the rest as on-orbit spares capable of immediate usage as primaries.

Name	Gen.	Launch	Manuf.	Location
TDRS-A	1	1983 Apr 04 / STS-6	TRW	49° W
TDRS-B	1	1986 Jan 28 / lost	TRW	—
TDRS-C	1	1988 Sep 29 / STS-26	TRW	275° W
TDRS-D	1	1989 Mar 31 / STS-29	TRW	41° W
TDRS-E	1	1991 Aug 2 / STS-43	TRW	174° W
TDRS-F	1	1993 Jan 13 / STS-54	TRW	47° W
TDRS-G	1	1995 Jul 13 / STS-70	TRW	150° W
TDRS-H	2	2000 Jun 30 / Atlas IIA	Boeing	171.5° W
TDRS-I	2	2002 Mar 08 / Atlas IIA	Boeing	170° W
TDRS-J	2	2002 Dec 04 / Atlas IIA	Boeing	150.7° W

The first seven satellites in the system (Generation 1) were manufactured by TRW and launched/deployed by shuttles. The three Generation 2 satellites were built by Boeing and launched on Atlas IIa boosters. The satellite summary is as follows:

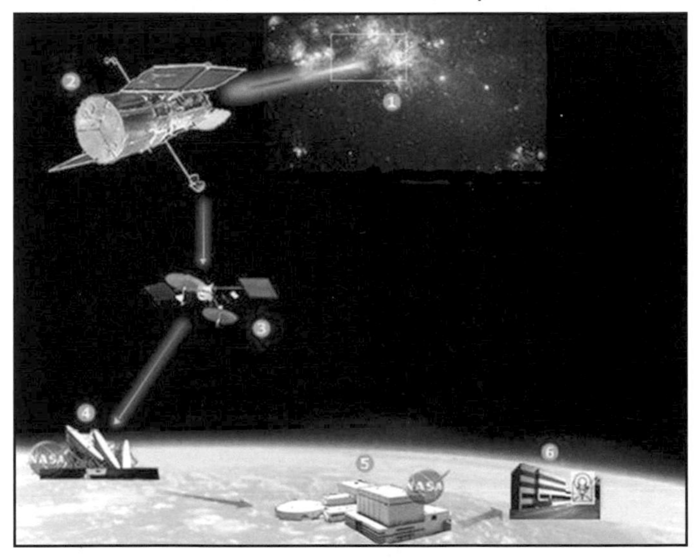

(1) A relatively nearby galaxy emits light that takes a long journey to Earth.
(2) Hubble Space Telescope, 380 miles above Earth – Hubble collects light from the target without interference from Earth's atmosphere. Hubble's instruments take a series of exposures through a variety of filters.
(3) Tracking and Data Relay Satellite System – A geosynchronous array of satellites receives data from Hubble about twice daily.
(4) White Sands Antenna Array – Ground stations exchange commands and data with the orbiting TDRSS. The White Sands station employs 60-foot-diameter high-gain microwave antennae.

(5) Space Telescope Operations Control Center (STOCC) – Through land lines and communications satellites the STOCC receives signals from White Sands. The STOCC is Hubble's "Mission Control." All commands and data go through this center.
(6) Space Telescope Science Institute (STScI) – The data is sent to the Institute for calibration and converted into the color images which most people see. Scientists here and all over world study these images and data.

Control Center

Hubble's flight controllers work in the Space Telescope Operations Control Center (STOCC) at NASA Goddard Space Flight Center in Greenbelt, Maryland. From there engineers and technicians "drive" the telescope 24 hours a day, seven days a week. Each of Hubble's four flight control teams consists of three operators backed by dozens of engineers and scientists. Controllers carry out routine operations in one section, while another section supports preparations for servicing missions, including testing and simulations. In an adjacent section, engineers perform in-depth subsystem analysis, conduct simulated subsystem tests, integrate new databases and validate new ground software and updates to flight software.

During Hubble Servicing

During a service mission, Hubble Mission Control becomes a much busier place. Extra teams of engineers monitor Hubble's vital signs as the astronauts install new instruments and make other improvements. Shortly after a service mission shuttle launch, the controllers command Hubble to capture attitude, configure it for shuttle rendezvous, close the aperture door and stow the high gain antennas.

Immediately after servicing, the controllers run tests on newly installed items, with the crew positioned at a safe location, to determine whether more astronaut activity is required. This is followed by ground-controlled detailed functional checkouts of the new equipment. After all of the servicing tasks are complete, Hubble is transferred back to internal power, its batteries are charged, and then the controllers reconfigure the telescope for normal science operations.

Scientific Instruments

Astronomy is the science that determines the precise positions and motions of stars and other celestial objects. These measurements are helping to advance knowledge of stars' distances, masses, and motions.

The Hubble Space Telescope's five science instruments – its cameras, spectrographs, and fine guidance sensors – work together and individually to record data that produces images from the farthest reaches of space. Each instrument was designed to observe the universe in a unique way.

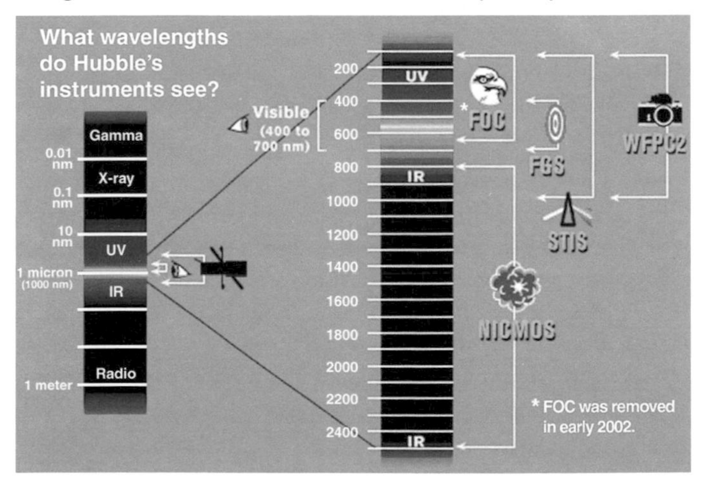

Wide Field and Planetary Camera

The Wide Field and Planetary Camera that was originally deployed with Hubble (WFPC1) was replaced during Service Mission 1 by WFPC2, an upgraded version of WFPC1 which includes corrective optics and improved detectors. WFPC2 is the instrument behind nearly all of Hubble most stunning pictures. As Hubble's main camera, it is used to observe just about everything, recording razor-sharp images of far away objects in relatively broad views. Although it is responsible for taking the images that most resemble human visual information, its 48 filters allow scientists to study precise wavelengths of light and to sense a range of wavelengths from ultraviolet to near-infrared light. WFPC2, like WFPC1 before it, is located in Hubble's radial instrument bay, which also houses the three fine guidance sensors.

WFPC2 doesn't use film to record its images. Instead, four postage stamp-sized pieces of high-tech circuitry called Charge-Coupled Devices (CCDs) collect the information on the telescope's focal plane to make photographs. These detectors are very sensitive to the extremely faint light of distant galaxies. They can see objects that are 1,000 million times fainter than the naked eye can see. Less sensitive CCDs are now in some videocassette recorders and all of the new digital cameras.

A CCD as an electronic array of light-sensitive picture elements (pixels), tiny cells that, placed together, resemble a mesh. Each of the four CCDs contains 640,000 pixels (800 x 800). The light collected by each pixel is translated into a number. These numbers (all 2,560,000 of them) are sent to ground-based computers, which convert them into an image.

Fine Guidance Sensors

The Fine Guidance Sensors (FGSs), in addition to providing Hubble with exceedingly accurate pointing capability, the FGSs can also be used to obtain highly accurate measurements of stellar positions. The three FGSs are installed in Hubble's radial instrument bay.

Goddard High Resolution Spectrograph

Goddard High Resolution Spectrograph (GHRS) was Hubble's first-generation spectrograph, used to obtain high-resolution spectra of bright targets. GHRS was removed from axial instrument bay 1 during Service Mission 2 when the Space Telescope Imaging Spectrograph was installed. GHRS successfully collected data for seven years.

GHRS was designed to detect ultraviolet light, which is emitted by all stars, planets, and galaxies. By looking at the ultraviolet light of stars, it can analyze the chemical composition and the gases that surround them, to help researchers understand the first moments of our universe.

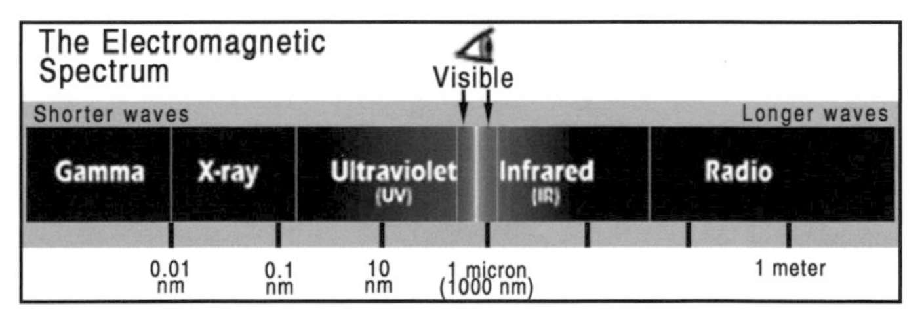

Faint Object Spectrograph

The Faint Object Spectrograph (FOS) was another Hubble first-generation spectrograph, used to obtain spectra of very faint or faraway sources. The FOS also had a polarimeter for the study of the polarized light from these sources. The FOS was designed to make spectroscopic observations of astrophysical sources from the near ultraviolet to the near infrared (1150 - 8000 Angstroms). The FOS was removed from axial instrument bay 2 during Service Mission 2 when the Near Infrared Camera and Multi-Object Spectrometer was installed.

Faint Object Camera

The Faint Object Camera (FOC) was a first-generation imaging camera used to image very small field-of-view, very faint targets. This was the last of the first-generation instruments to be removed from Hubble. It was Hubble's telephoto lens for nearly 12 years, until it was removed from axial instrument bay 3 during Service Mission 3B when the Advanced Camera for Surveys was installed.

High Speed Photometer

The High Speed Photometer (HSP) was Hubble's first-generation photometer, used to measure very fast photometric observations (brightness changes) in diverse astrophysical sources in a variety of filters and passbands, from the near ultraviolet to the visible. The HSP was removed from axial instrument bay 4 during Service Mission 1 to make room for the Corrective Optics Space Telescope Axial Replacement package (COSTAR).

COSTAR

The Corrective Optics Space Telescope Axial Replacement package (COSTAR) is not an actual instrument. It is a complex packaging of five optical mirror pairs which refocused the light from Hubble's optical system for first-generation instruments, correcting for the primary mirror's spherical aberration, discovered after Hubble deployment. There are no first-generation instruments still in operation on Hubble, so COSTAR is no longer in use. It is still installed in axial instrument bay 4.

Space Telescope Imaging Spectrograph

The Space Telescope Imaging Spectrograph (STIS), separating light into component wavelengths, much like a prism, provided scientists with spectra and images at ultraviolet and visible wavelengths, probing the universe from our solar system out to cosmological distances.

STIS is a second-generation imager/spectrograph that can obtain high resolution spectra from many different points along a target, giving data about their temperature, chemical composition, density and motion. STIS spans ultraviolet, visible, and near-infrared wavelengths. STIS replaced GHRS, both functionally and physically, in axial instrument bay 1 during Service Mission 2.

Near Infrared Camera and Multi-Object Spectrometer

The Near Infrared Camera and Multi-Object Spectrometer (NICMOS), a second-generation imager/spectrograph, is Hubble's only near-infrared (NIR) instrument. As Hubble's "heat sensor," NICMOS can see through interstellar gas and dust to objects in deepest space whose light takes billions of years to reach Earth.

Just as a camera for recording visible light must be dark inside to avoid exposure to unwanted light, a camera for recording infrared light must be cold inside to avoid exposure to unwanted light in the form of heat, therefore NICMOS must operate at very cold temperatures, below –321 degrees Fahrenheit.

NICMOS was installed in axial instrument bay 2 during Service Mission 2, replacing the Faint Object Spectrograph. NICMOS was turned off during Service Mission 3A because of problems with its cooling system. It was turned back on during Service Mission 3B after the NICMOS Cooling System was installed.

NICMOS Cooling System

The NICMOS Cooling System (NCS) is not a separate instrument but rather a device that allowed NICMOS to resume operation by providing mechanical cooling for the NICMOS detectors. Estimates are that the NCS will allow NICMOS to operate until about 2007.

Advanced Camera for Surveys

The Advanced Camera for Surveys (ACS) is a third-generation Hubble instrument that, among other tasks, will be used to observe weather on other planets in our solar system, conduct new surveys of the universe, and study the nature and distribution of galaxies.

With a wider field of view and better light sensitivity, ACS effectively increases Hubble's discovery power by 10 times. ACS outperforms all previous instruments flown aboard Hubble, primarily because of its expanded wavelength range. Designed to study some of the earliest activity in the universe, ACS sees in wavelengths ranging from far ultraviolet to infrared.

ACS was installed during Servicing Mission 3B in axial instrument bay 3, replacing the Faint Object Camera as Hubble's "zoom lens."

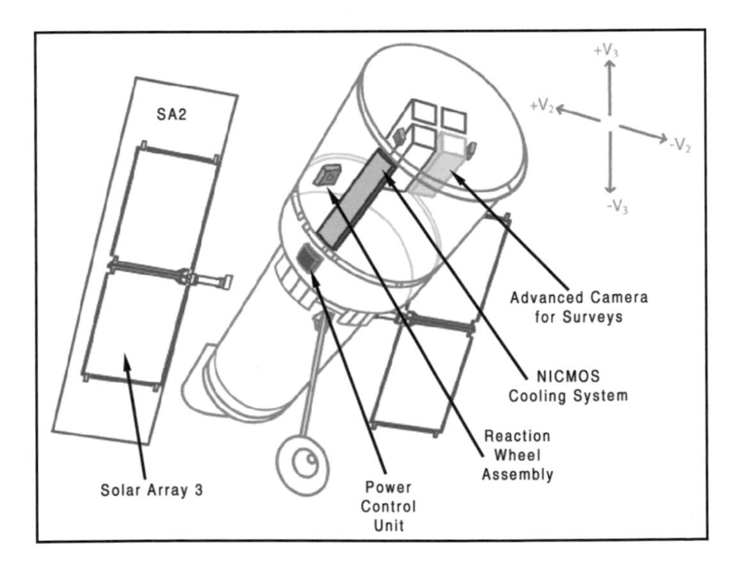

Wide Field Camera 3

The Wide Field Camera 3 (WFC3) is a fourth-generation imaging camera that will supplement ACS if installed. It is slated to be installed during the possible Servicing Mission 4.

Cosmic Origins Spectrograph

The Cosmic Origins Spectrograph (COS) is a fourth-generation spectrometer. It is an ultraviolet spectrograph optimized for observing faint point sources with moderate spectral resolution. It is slated to be installed during the possible Servicing Mission 4.

The Science of the Hubble Telescope

Capturing Images:

To understand what Hubble sees, it is necessary to understand what astronomical telescopes do, and how they differ from the ones on the ground. All telescopes collect light and magnify it. When an image is magnified, it gets dimmer. This is because the light coming from the object is spread out over a larger area. If we try to magnify the image too much, it gets so dim when we look through the telescope we don't see anything at all.

Astronomers have come up with several creative ways to obtain more light and make their images brighter. One way astronomers get around this problem is to make the telescope larger. A larger telescope will collect more light from the object, so the image is brighter too.

Another way is to use a recording device instead of the human eye. Recording devices can be photographic films or electronic detectors. Unlike the human eye, these methods have an advantage in that exposures of faint objects can made over a very long time. The exposures can be as long as 45 minutes per image. The images can then be added together to add up to exposure times of up to 38 hours for each filter. The longer an exposure is, the more light falls on the film. As long as the image is held very still, eventually an image of the faint target will appear. Another advantage of recording devices is that a permanent copy of the observation is obtained.

Hubble looks at objects which are so distant and faint, that even with highly sensitive CCD detectors (which are many times more sensitive than the human eye) it must expose for long periods of time. These exposures can be for 30 minutes or more. Sometimes hundreds of half-hour exposures are added together to produce one very long exposure. Observations using CCD detectors can also be added together, which is something that cannot be done with photographic film.

Color Image Processing

The Hubble Space Telescope provides amazing color pictures of planets, stars, galaxies and nebulae – and does so without the use of a color camera. The telescope's cameras record light as grayscale images – that is, black-and-white images with varying shades of gray, like an old black-and-white movie. Two or more of these black-and-white images are combined during the picture processing to create a color picture.

Three types of color picture can be produced from the grayscale images recorded by Hubble: Natural Color, Enhanced Color and Representative Color.

A natural color picture is a "normal" view, in which the "real" colors are used; the colors are as they would appear to your eye if you were close enough to see them. A natural color picture (see page 52) is produced by combining a red image, a green image and a blue image, the same as is done to produce a color television screen. The three images are created by recording the same view three times using a red, green and then blue filter, respectively.

An enhanced color picture is produced in the same manner as a natural color picture, except that instead of using the normal red, green and blue filters corresponding to the visible colors, filters are selected specifically for visible colors of particular interest. For example, the Cat's Eye Nebula (see page 53) ejects hydrogen atoms, oxygen atoms and nitrogen ions. The wavelengths corresponding to these three elements are all in the red band, close enough together that they would not be separately discernible in a natural color picture. The picture on page 53 was produced by combining three images, each using a filter for one of the three specific red wavelengths. What results is picture where light from hydrogen atoms is shown in red, light from oxygen is shown in blue, and light from nitrogen is shown in green, thereby clearly showing information that would not show in a natural color picture.

In a representative color picture, red, green and blue are assigned to images made with filters for non-visible wavelengths (infrared and ultraviolet). This produces a picture in which information about the non-visible wavelengths is easily discernible. For example, in the representative color picture of Saturn on page 54 information about infrared radiation is clearly visible. Blue has been assigned to the shortest-wavelength infrared light, red to the longest-wavelength infrared light, and green to the intermediate-wavelength infrared light (there is no visible light displayed in this picture). The colorful bands arise because chemical differences in Saturn's upper cloud layers cause those clouds to reflect sunlight in different ways.

Hubble Image Mosaics

Many Hubble images have a stair-step shape. These images are produced by the Wide Field and Planetary Camera 2 (WFPC2). WFPC2 has four cameras, each of which records one quadrant of the total view. One of the four cameras records a magnified view, which produces finer detail than the other three. When processing combines the four quadrants into a single image, the magnified quadrant is reduced to the same scale as the other others, leaving the rest of that quadrant blank, since there is no data for it.

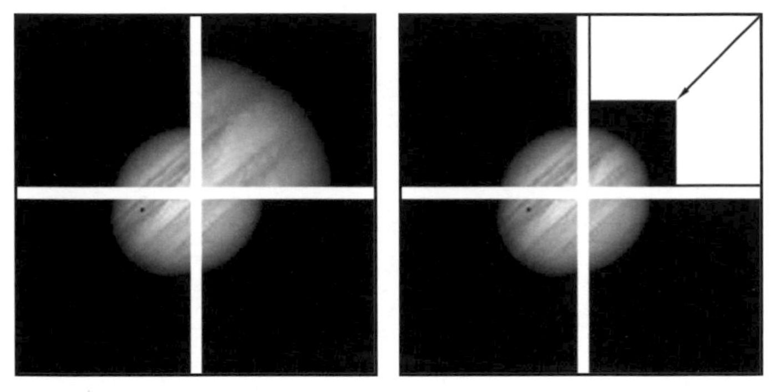

Example of how WFPC2 camera #1 (high-resolution quadrant) is reduced to match other quadrants, resulting in the stair-step photograph.

Stars

The life time of a star is millions or billions of years. Since we can't see changes in individual stars, we must piece together snapshots of different stars at different stages of their lives. The birth, life, and rebirth of stars is an ongoing process, byproducts of which include planets and the elements necessary to create life.

Stars burn hydrogen and helium (fusion) over the course of their lives. A star's temperature and color are in direct proportion to its rate of burning. A hot blue star lives only a few million years. Yellow stars, like our Sun, live for about 10 billion years, and cooler red stars live for tens of billions of years. Most stars simply exhaust their fuel and then die peacefully, while other, hotter explode as novae or supernovae.

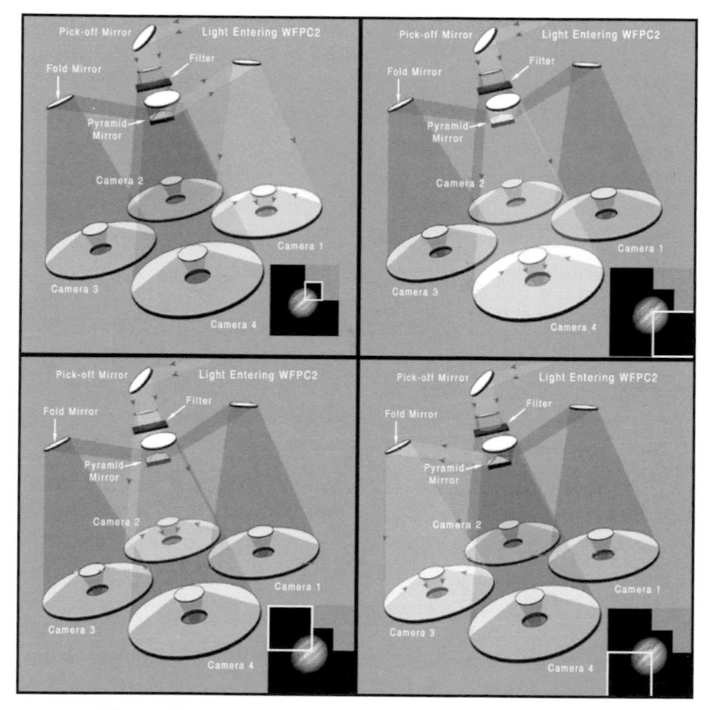

Illustration of light paths for the four WFPC2 cameras.

Star Birth

Hubble has captured vivid views of newly forming stars inside the Orion nebula. Stars form when a cloud of interstellar gas collapses under the force of its own gravity. Because the cloud is spinning slowly, It leaves behind a dark disk of orbiting debris that may well form a planetary system.

About 4.6 billion years ago, before planets formed around our Sun, our solar system was probably a dark disk of debris surrounding the central sun. The matter in these disks gradually collected into ever larger lumps that ultimately became planet-sized.

As a spinning, collapsing cloud enters the final phase before becoming a full-fledged star, violent things happen. Hubble has recorded spectacular jets of high-speed gas shooting out from the pole of a forming star.

Star Death

After a star like our Sun has used up all of its nuclear fuel, it throws off its outer layers and then fades away. Hubble's images of these glorious displays (called planetary nebulae) have provided new insights into the complexities of a star's final days. Despite the name, planetary nebulae have little to do with planets. They were named over a century ago because through the small telescopes of the day they resembled planets.

About Galaxies

Galaxies come in diverse shapes and sizes. In a spiral galaxy, like the Milky Way, central region contains primarily older, yellow and red stars. The outer spiral arms are considerably typically bluer, due to ongoing formation of young, blue stars. The stars form a flat disk that circles the galactic nucleus.

Galactic Black Holes

Most galaxies (if not all) have immense black holes at their centers. We can't see them, but instruments like the spectrograph on Hubble can measure how the black holes affect the matter surrounding them, which is how we determine their presence.

The Universe

Hubble's longest exposures are like a core sample of the universe, recording galaxies at many different distances. Massive objects such as clusters of galaxies bend space in their vicinity, distorting and in some cases magnifying the objects behind them. This interesting effect is called gravitational lensing.

Gravitational lenses are being used to study young galaxies and to gain a better understanding of the total amount of matter in the universe. Hubble's observations of gravitational lenses show that the lensing clusters must have more matter than meets the eye.

Spectrographs

A spectrograph is an instrument for measuring light intensity at a specific wavelength (color) or wavelength rage. By selecting very specific wavelength ranges, scientists could look for specific features, such as the hottest stars in a particular cluster.

Using the Faint Object Camera (FOC, built by the European Space Agency) and later the Advanced Camera for Surveys (ACS) Hubble scientists could produce pictures of very faint objects by recording images over long exposure times and combining exposures. The total image was converted into digital data, transmitted to Earth, and then reconstructed.

Hubble's instruments have the ability to single out individual stars in distant star clusters. Since Hubble can make high-resolution observations of faint sources at ultraviolet and visible wavelengths, it can study star clusters, examine galaxies and faint objects (such as quasars), and look for small details of celestial objects.

Advanced Camera for Surveys

Among other tasks, ACS maps the distribution of dark matter, detects the most distant objects in the universe, searches for massive planets, and studies the evolution of clusters of galaxies.

ACS has three different cameras, each designed to perform a specific function. The ACS's wide-field camera conducts broad surveys of the universe. Astronomers use it to study the nature and distribution of galaxies, which reveal clues about how our universe evolved. The high-resolution camera takes extremely detailed pictures of the inner regions of galaxies. It searches neighboring stars for planets and protoplanets, and takes close-up images of the planets in our own solar system.

The solar blind camera, which blocks visible light to enhance ultraviolet sensitivity, focuses on hot stars radiating in ultraviolet wavelengths.

Near Infrared Camera and Multi-Object Spectrometer

Many secrets about the birth of stars, solar systems, and galaxies are revealed in infrared light, which can penetrate the interstellar gas and dust that block visible light. Also, light from the most distant objects in the universe doppler shifts into the infrared wavelengths. By studying objects and phenomena in the infrared region, astronomers can probe our universe's past, present and future, learn how galaxies, stars, and planetary systems form, and reveal a great deal about our universe's basic nature.

Space Telescope Imaging Spectrograph

On August 3, 2004, STIS stopped science operations and is currently in "safe" mode. In addition to taking detailed pictures of celestial objects, the versatile STIS acts like a prism to separate light from the cosmos into its component colors. This provides a wavelength "fingerprint" of the object being observed, which gives data about its temperature, chemical composition, density, and motion. Spectrographic observations also reveal changes in celestial objects as the universe evolves. STIS spans ultraviolet, visible, and near-infrared wavelengths.

Astronomers used STIS to hunt for black holes. The light emitted by stars and gas orbiting the center of a galaxy appears redder when moving away from us (redshift), and bluer when coming toward us (blueshift). STIS is looking for redshifted material on one side of the suspected black hole and blueshifted material on the other, indicating that this material is orbiting an object at very high speeds.

STIS can sample 500 points along a celestial object simultaneously. This means that many regions in a planet's atmosphere or many stars within a galaxy can be recorded in one exposure, vastly improving Hubble's speed and efficiency.

Hubble Discoveries

The Hubble Space Telescope has had a major impact in every area of astronomy, from the solar system to objects at the edge of the universe. To date, more than 3,500 technical publications have reported HST results. Following is a summary of Hubble's major scientific results.

The accelerating universe and dark energy – Hubble's ability to detect faint supernovae contributed to the discovery that the expansion rate of the universe is accelerating, indicating the existence of mysterious "dark energy" in space.

The distance scale and age of the universe – Observations of Cepheid variable stars in nearby galaxies were used to establish the expansion rate of the universe to better than 10 percent accuracy.

The evolution of galaxies – The Hubble Deep Field provided our deepest view yet into the universe's distant past, allowing us to reconstruct how galaxies evolve and grow by swallowing other galaxies.

The birth of stars and planets – Peering into nearby regions of star birth in the Milky Way galaxy, Hubble has revealed flattened disks of gas and dust that are the likely the birthplaces of new planets.

Stellar death – When Sun-like stars end their lives, they eject spectacular nebulae. Hubble has revealed fantastic and enigmatic details of this process.

Stellar populations in nearby galaxies – Deep images that resolve individual stars in other galaxies reveal the history of star formation.

Planets around other stars – Hubble made detailed measurements of a Jupiter-sized planet orbiting a nearby star, including the first detection of the atmosphere of an extrasolar planet.

The impact of comet Shoemaker-Levy 9 on Jupiter – The explosive collision of the comet with Jupiter provided a cautionary tale of the danger posed by cometary impacts.

Black holes in galaxies – Hubble observations have shown that monster black holes, with masses millions to billions times the mass of our Sun, inhabit the centers of most galaxies.

Gamma-ray bursts – Hubble played a key role in determining the distances and energies of gamma-ray bursts, showing that they are the most powerful explosions in the universe other than the big bang itself.

Galaxy Evolution

Hubble has peered across space and time to study galaxies in an infant universe. The most famous of Hubble's faraway views is the Hubble Deep Field, a tiny speck of sky that revealed a zoo of about 3,000 galaxies, some as old as 10 billion years. The Hubble Deep Field, taken in 1995, has become one of the most studied regions of the sky and has been examined in a wide range of wavelengths, from radio to infrared.

Hubble's observations of deep space indicate that the young cosmos was filled with much smaller and more irregularly shaped galaxies than those that astronomers see in our nearby universe. These smaller structures, composed of gas and young stars, may be the building blocks from which the more familiar spiral and elliptical galaxies formed, possibly through processes such as multiple galaxy collisions and mergers.

A Speedy Universe

In 2001, Hubble identified the farthest stellar explosion to date, a supernova that erupted 10 billion years ago. By examining the glow from this dying star, a supernova called 1997ff, astronomers collected the first observational evidence that gravity began slowing down the universe's expansion after the Big Bang.

This reinforces the idea that the universe only recently began speeding up, a discovery made in 1998 when the unusually dim light of several distant supernovas suggested that the universe is expanding more quickly than it did in the past. Many scientists believe that a mysterious, repulsive force is at work in the cosmos, making galaxies rush ever faster away from each other.

Age of the Universe

The universe has been expanding since its creation in the Big Bang. Astronomer Edwin Hubble made that observation in the 1920s. Since then, astronomers have debated how fast the cosmos is expanding, a value called the Hubble constant. In May 1999 a team of astronomers announced that they had obtained a value for the Hubble constant, an essential ingredient needed to determine the age, size, and fate of the universe. They did it by measuring the distances to 18 galaxies, some as far as 65 million light-years from Earth. After obtaining a value for the Hubble constant, the team then determined that the universe is 12 to 14 billion years old. Measuring the Hubble constant was one of the three major goals for NASA's Hubble Space Telescope before it was launched in 1990.

In April 2002, another team of astronomers announced that they had used a different age-dating technique to reach a similar estimate for the universe's age: between 12 and 13 billion years. The team based their estimate on Hubble telescope observations of the oldest and faintest burned-out stars, called white dwarfs, in the Milky Way Galaxy. These extremely old, dim stars provide a completely independent reading on the age of the universe without relying on measurements of the expansion rate of the universe.

Black Hole Hunter

Hubble also yielded clues to what is causing the flurry of activity in the hearts of many galaxies. These central regions are very crowded, with stars, dust, and gas competing for space. But Hubble managed to probe these dense regions, and in 1994 the telescope provided decisive spectroscopic evidence that supermassive black holes exist. Supermassive black holes are compact "monsters" that are millions or billions times more massive than our Sun and gobble up any material that ventures near them. These elusive "eating machines" cannot be observed directly, because nothing, not even light, escapes their stranglehold.

But the telescope did capture dramatic photographs of quasars, energetic light beacons that astronomers believe are powered by black holes. These photographs, released in 1996, revealed that quasars live in a variety of galaxies, from normal spiral galaxies to distorted colliding galaxies.

In 1997, a Hubble census of 27 nearby galaxies showed that supermassive black holes are common in large galaxies. The census also revealed a relationship between a black hole's mass and the mass of its home galaxy.

After proving that black holes are ubiquitous, the orbiting observatory then began further examining the relationship between supermassive black holes and their home galaxies. In 2000, a census of more than 30 galaxies showed that a galaxy's bulge determines the mass of its black hole.

Planet-Making Recipe

The Hubble telescope provided visual proof that pancake-shaped dust disks around young stars are common, suggesting that the raw material for planet formation is in place. In 1994, the telescope revealed that these disks are swirling around at least half of the stars in the Orion Nebula, a cauldron of star formation. The finding reinforces the assumption that planetary systems are common in the universe. Scientists believe that the Earth and other planets of the solar system were formed out of similar disks about 4.5 billion years ago by the coalescing of matter caused by gravitational attraction.

In 2001, astronomers using the Hubble telescope made the first direct detection of the atmosphere of a planet orbiting a star outside our solar system and obtained the first information about the planet's chemical composition. The planet, a gas giant like Jupiter, orbits a Sun-like star called HD 209458, located 150 light-years away in the constellation Pegasus. The orbiting observatory probed the planet's atmospheric composition by watching it pass in front of its parent star, allowing astronomers for the first time ever to see light from the star filtered through the planet's atmosphere. Astronomers then analyzed the light to determine the type of gases present in the planet's atmosphere.

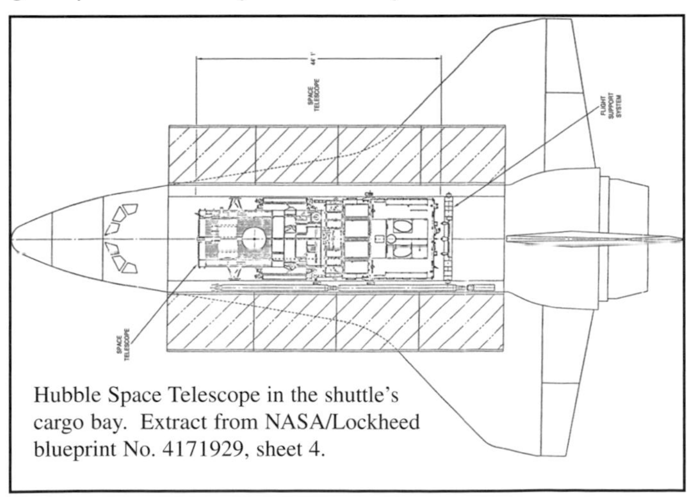

Hubble Space Telescope in the shuttle's cargo bay. Extract from NASA/Lockheed blueprint No. 4171929, sheet 4.

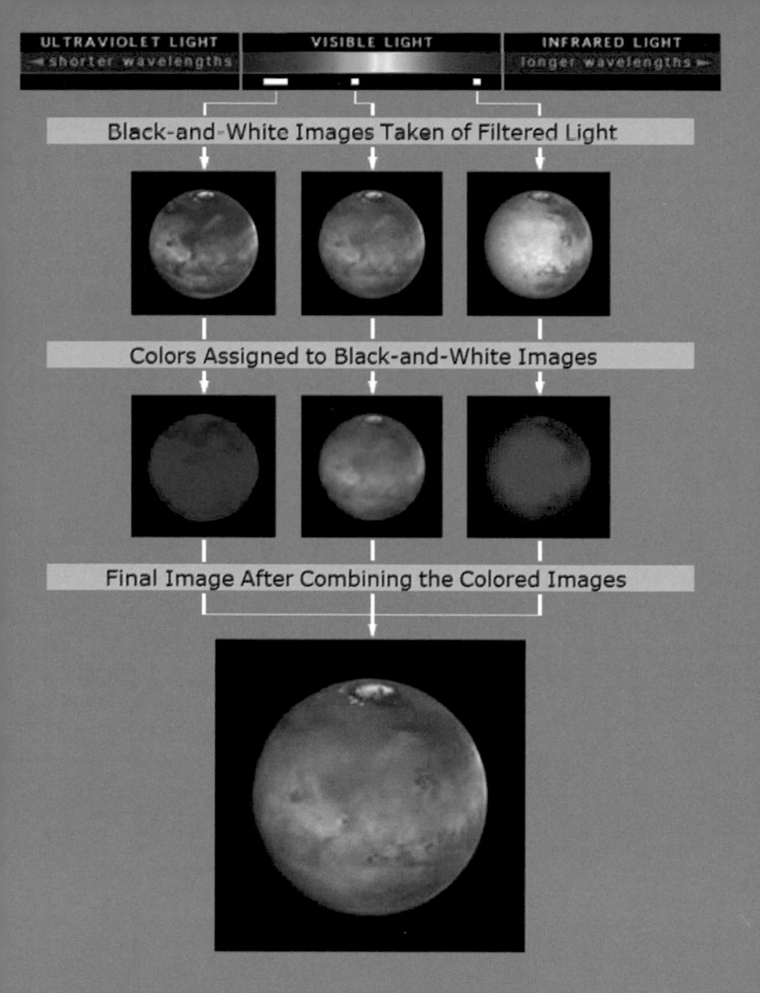

A natural color picture is a "normal" view, in which the "real" colors are used; the colors are as they would appear to your eye. Three images of the view are recorded, one each through a red filter, a green filter and a blue filter, and the three images are then combined to create a color photograph, the same as for a color television screen.

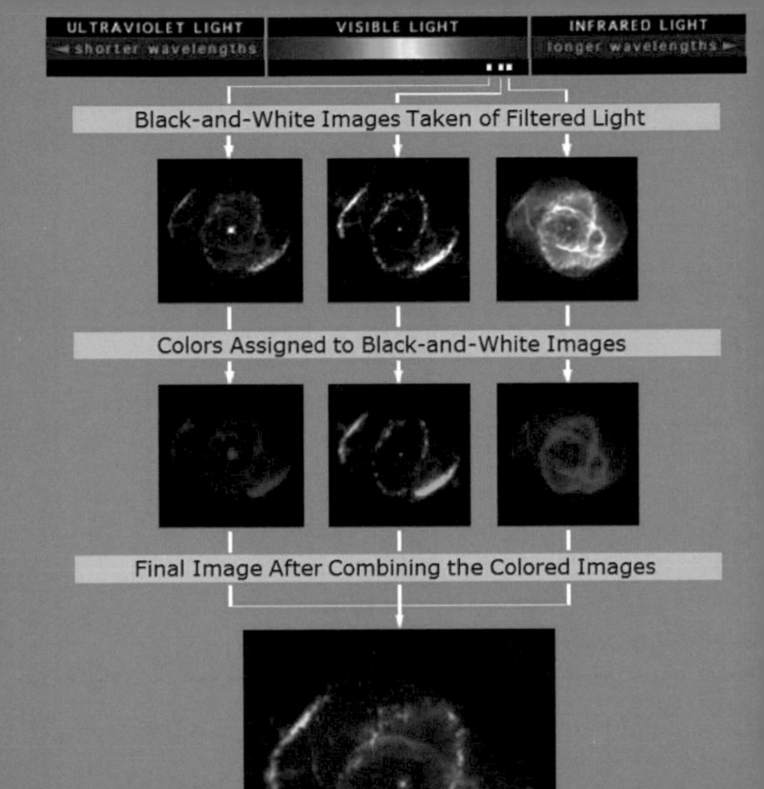

An enhanced color picture is produced in the same manner as a natural color picture, except that instead of the normal red, green and blue filters, filters are selected specifically for visible wavelengths of particular interest. In the above enhanced picture, the filters are for the infrared wavelengths of hydrogen, oxygen and nitrogen atoms, which are then assigned to the red, blue and green channels, respectively. The resulting picture therefore shows the densities the density of the three elements within the field of view, and looks nothing like the natural color picture of the same view.

A representative color picture, is similar to an enhanced color picture except that it is made with filters for non-visible wavelengths (infrared or ultraviolet). This produces a picture in which information about the non-visible wavelengths is easily discernible. In the example above, the representative colors are for infrared wavelengths and therefore clearly give thermal information about the planetary surface. Again, it looks nothing like the natural color picture of the same view.

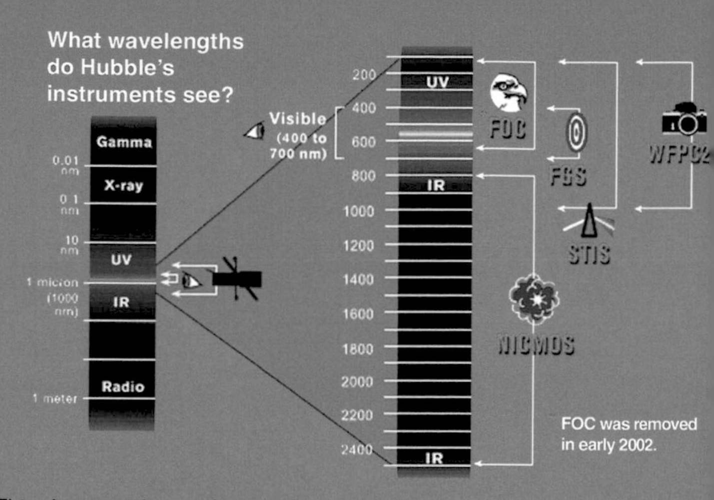

This diagram shows the wavelength range within the electromagnetic spectrum that Hubble's scientific instruments can record, and which portion of that range each instrument can record.

Many Hubble images have a stair-step shape. These images are produced by Wide Field and Planetary Camera 2, which has four cameras, one for each quadrant of the total view. One of the four cameras records a magnified view, which produces finer detail than the other three. When processing combines the four quadrants into a single image, the magnified quadrant is reduced to the same scale as the other others, leaving the rest of that quadrant blank, since there is no data for it.

Delivering the Detail

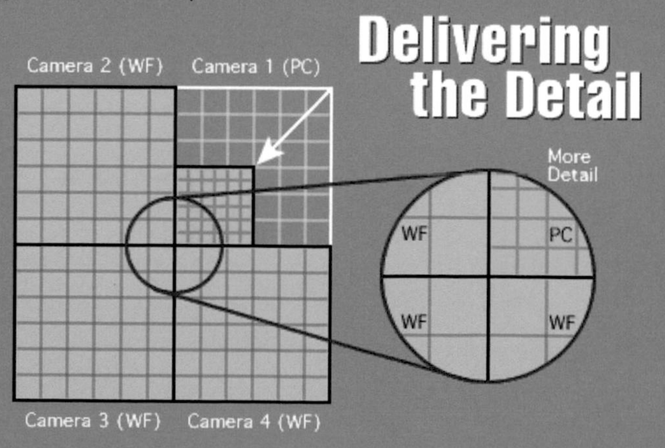

As Hubble's third repair mission unwinds, ground controllers send commands and instructions to the telescope while monitoring data from the observatory to check that it continues to function correctly.

Engineers in a clean room at Ball Aerospace in Boulder, Colorado, work on one of Hubble's instruments, the Space Telescope Imaging Spectrograph (STIS), in 1996. The instrument, installed in Hubble in 1997, breaks light into colors, giving scientists an important analytical tool for studying the cosmos.

Astronauts train to service Hubble in a huge, water-filled tank that simulates weightlessness. The 40-foot-deep (12 m) tank at NASA's Johnson Space Center contains full-scale underwater mock-ups of Hubble, its instruments, and the carriers that hold the instruments.

Carriers, used to transport Hubble instruments, wait in the largest clean room in the world, Goddard Space Flight Center's 1.3-million-cubic-foot (0.036 million cubic meters) High Bay Clean Room.

Hubble is lifted into the upright position at Kennedy Space Center in preparation for its 1990 launch aboard the Space Shuttle Discovery. A closer look at this image reveals a portion of the 225 feet (68.6 meters) handrails installed around the outside for astronauts to grip during repair mission spacewalks.

The Space Shuttle Endeavour is being rolled around from Launch Pad 39A to Launch Pad 39B. The rare pad switch was deemed necessary after contamination was discovered in the Payload Change-out Room at Pad A. Still to come are the payloads for the upcoming STS-61 mission, the first servicing of the Hubble Space Telescope.

Workers study Hubble's main mirror. When light enters Hubble, it reflects off the main mirror and strikes a second, smaller mirror, beyond which Hubble's science instruments wait to capture it. In this photo, the hole in the mirror's center is covered up.

Astronauts train in a Goddard Space Flight Center clean room on a backup of an electrical section of Hubble in 1999. The astronauts wear "bunny suits," special coveralls, hoods, gloves, boots and masks that protect the sensitive equipment from particles that could interfere with its performance.

The image shows two astronauts practicing construction techniques to build Space Station in Neutral Buoyancy Simulator (NBS) at Marshall Space Center (MSFC) in 1985 to evaluate techniques that were used in space to assemble structures such as the Hubble Space Telescope.

In the Vehicle Assembly Building a NASA Quality Control inspector checks the connection between Space Shuttle *Discovery* and the external tank that was used to launch mission STS-103, during which the crew replaced aging parts on the nine year old Hubble Space Telescope.

The Space Shuttle *Columbia* on Pad 39A during the picture-perfect ascent of sister ship *Discovery* after lift off of STS-31.

The Space Shuttle *Discovery* takes off on a mission to upgrade and repair Hubble in 1999. Once it gets close enough to Hubble, the shuttle uses its robotic arm to tow the telescope into its cargo bay for astronauts to work on. Astronauts routinely visit Hubble to perform maintenance work and install new instruments, thanks to the telescope's unique construction with replaceable parts.

An astronaut works on Hubble with a ratchet during the second servicing mission in 1997. NASA specially designed the power tool to withstand the harsh environment of space, making it an essential item during three different Hubble missions.

An astronaut during Hubble's second servicing mission in 1997 prepares to document the day's activities with a shuttle camera. Engineers rely on astronauts' photos to design and build new hardware for Hubble, and other astronauts use them for training.

The Hubble Space Telescope rests in the Space Shuttle *Discovery*'s cargo bay during the third repair mission in December 1999. Hubble must attach to the shuttle for astronauts to perform repairs. *Discovery* is the shuttle that originally carried Hubble into orbit in 1990.

An astronaut anchored on the end of the Remote Manipulator System (RMS) arm prepares to be elevated to the top of the Hubble Space Telescope (HST) to install protective covers on the magnetometers.

Astronauts headed for Hubble during its second servicing mission in 1997 expected to repair some of the telescope's outside insulation, which deteriorates in the harsh environment of space. But when they arrived, they discovered more damage than they had expected. Working inside the shuttle, they made patches out of materials in their repair kit.

Astronauts replace the Hubble Space Telescope's Fine Guidance Sensors (FGS) in 1997. The FGS are used to locate and lock onto a target star while science instruments make observations. They can also perform measurements on the positions and motions of stars.

Attached to the "robot arm" the Hubble Space Telescope is unberthed and lifted up into the sunlight during the second servicing mission.

Astronauts remove the Wide Field and Planetary Camera to replace it with its more powerful successor, Wide Field and Planetary Camera 2, during Hubble's first servicing mission in 1993.

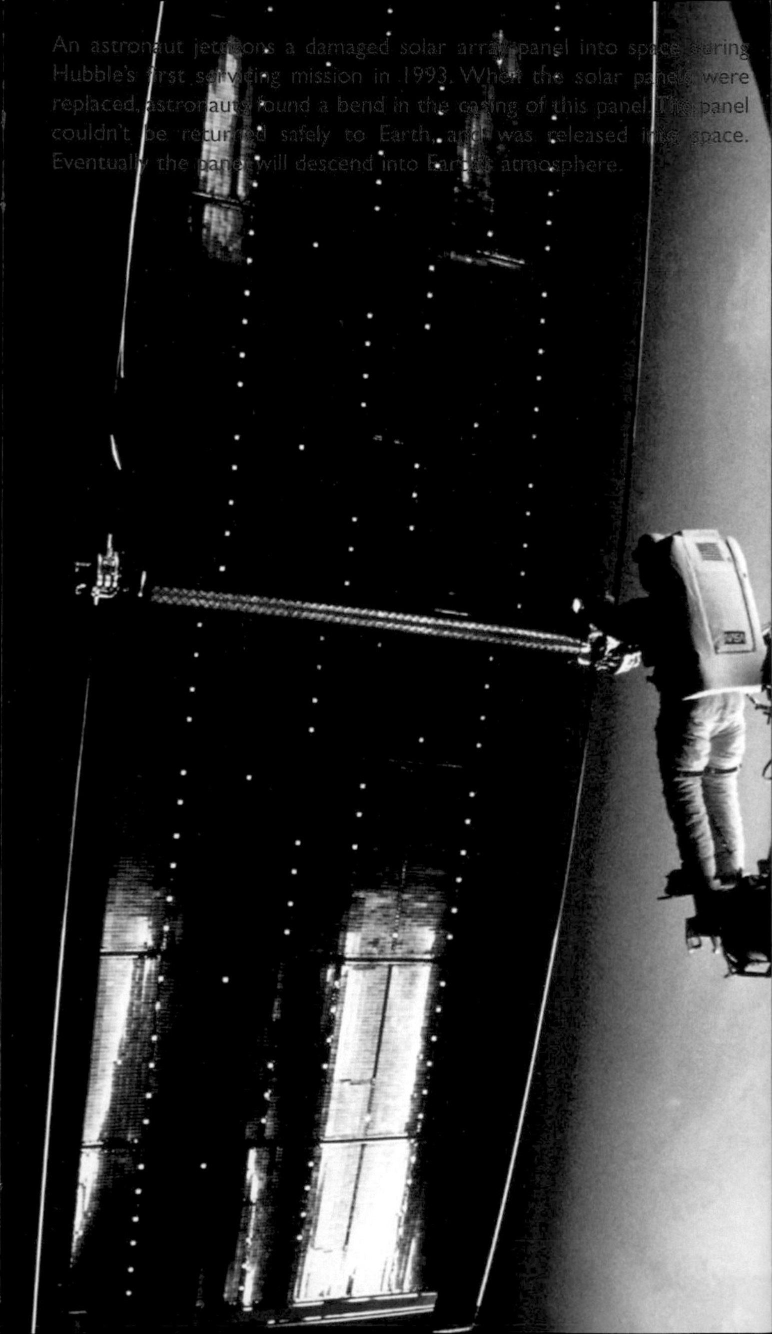

An astronaut jettisons a damaged solar array panel into space during Hubble's first servicing mission in 1993. When the solar panels were replaced, astronauts found a bend in the casing of this panel. The panel couldn't be returned safely to Earth, and it was released into space. Eventually the panel will descend into Earth's atmosphere.

An astronaut removes the High Resolution Spectrograph in preparation for a new instrument during the second servicing mission in 1997. Hubble's science instruments are large and complex. The telescope can hold four telephone-booth sized instruments and four piano-sized instruments.

Astronauts conducting Detailed Test Objective (DTO) procedures in the payload bay of *Endeavour*, results of which assisted in refining several procedures developed to service the Hubble Space Telescope (HST) on mission STS-61.

An astronaut conducts an in-space evaluation of the Portable Foot Restraint (PFR) which was later used on the first Hubble Space Telescope (HST) servicing mission. He is positioned on the edge of *Discovery*'s payload bay.

The Hubble Space Telescope hovers at the boundary of Earth and space in this picture, taken after Hubble's second servicing mission in 1997. Hubble drifts 353 miles (569 km) above the Earth's surface, where it can avoid the atmosphere and clearly see objects in space.

Backdropped against the blue and white Earth, astronauts wearing Extravehicular Mobility Units (EMUs), simulate handling of large components in space.

The Hubble Space Telescope drifts through space (below) in this picture, taken by Space Shuttle *Discovery* during Hubble's second servicing mission in 1997. The 10-foot aperture door, open to admit light, closes to block out space debris.

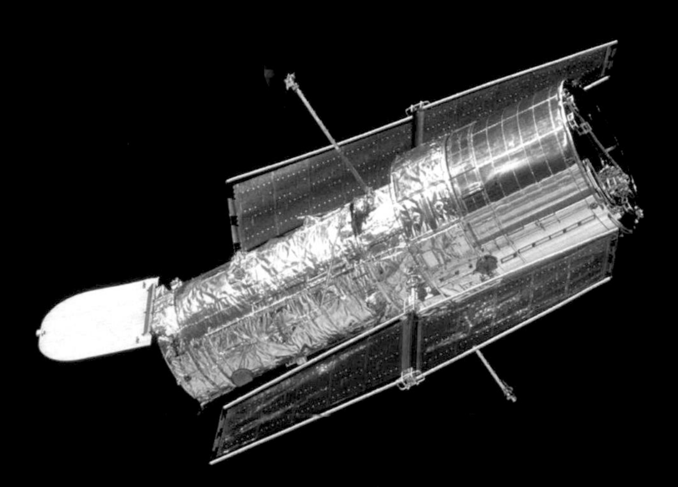

The Hubble Space Telescope floats against the background of Earth after a week of repair and upgrade by Space Shuttle Columbia astronauts in 2002. Hubble's fourth servicing mission gave the telescope its first new instrument installed since the 1997 repair mission – the Advanced Camera for Surveys. It doubled Hubble's field of view and records information much faster than Hubble's Wide Field and Planetary Camera 2.

Taking advantage of Mars's closest approach to Earth in eight years, astronomers using NASA's Hubble Space Telescope have taken the space-based observatory's sharpest views yet of the Red Planet. The telescope's Wide Field and Planetary Camera 2 snapped these images when Mars was 54 million miles (87 million kilometers) from Earth.

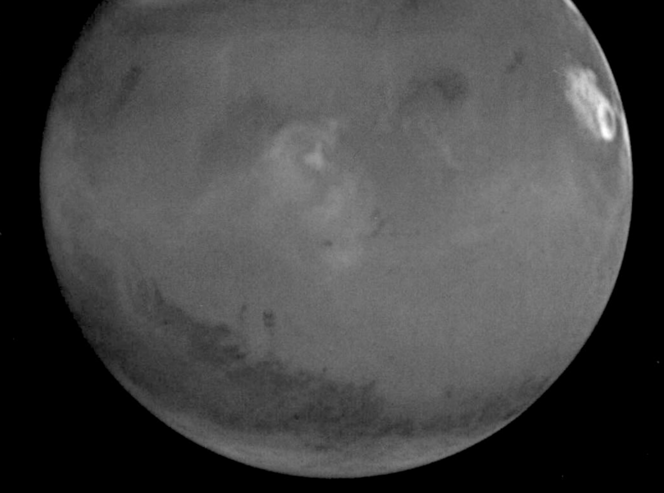

The top image is centered near a volcanic region known as Elysium. Along the right limb, a large cloud system has formed around the Olympus Mons volcano. The bottom image is centered on the region of the planet known as Tharsis, home of the largest volcanoes in the solar system. Prominent late afternoon clouds along the right limb of the planet can be seen.

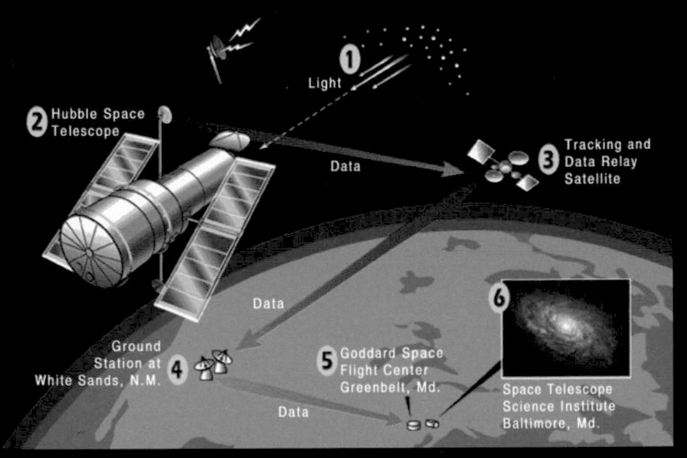

This visual summary of the Hubble data flow shows how a scene billions of light-years away becomes photographs on Earth. The Tracking Data and Relay Satellite System is used by many projects besides Hubble.

This picture of Neptune was produced from the last whole planet images taken through the green and orange filters on the *Voyager 2* narrow angle camera. Years later, when the Hubble telescope was focused on the planet, these atmospheric features had changed, indicating that Neptune's atmosphere is dynamic.

Located some 13 million light-years from Earth, NGC 4214 is currently forming clusters of new stars from its interstellar gas and dust. In this Hubble image, we can see a sequence of steps in the formation and evolution of stars and star clusters.

Jupiter's Great Red Spot changes its shape, size, and color. These changes are demonstrated in high-resolution Wide Field and Planetary Cameras 1 & 2 images of Jupiter obtained by the Hubble Space Telescope. With a diameter of 15,400 miles, the Red Spot is the largest known storm in the Solar System.

Astronomers, using the Wide Field Planetary Camera 2 on board NASA's Hubble Space Telescope in October and November 1997 and April 1999, imaged the Bubble Nebula (NGC 7635) with unprecedented clarity.

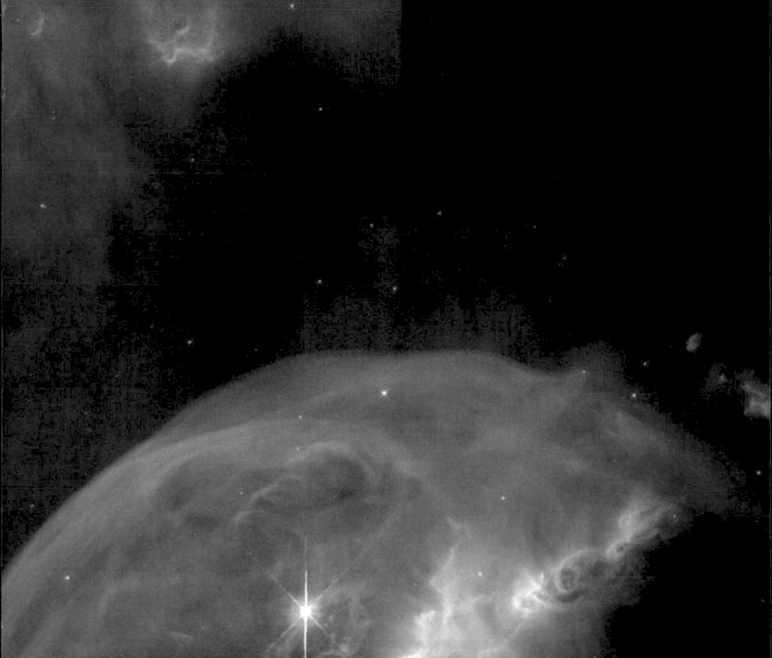

Previously unseen details of a mysterious, complex structure within the Carina Nebula (NGC 3372) are revealed by this Hubble image (below) of the "Keyhole Nebula." The picture is a montage assembled using six different color filters. This region, about 8000 light-years from Earth.

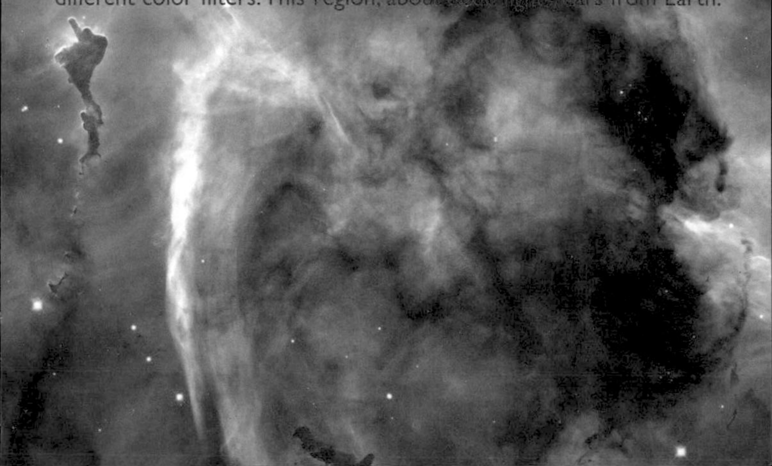

Hubble imaged this giant, cosmic magnifying glass, a massive cluster of galaxies called Abell 2218. This 'hefty' cluster resides in the constellation Draco, some 2 billion light-years from Earth.

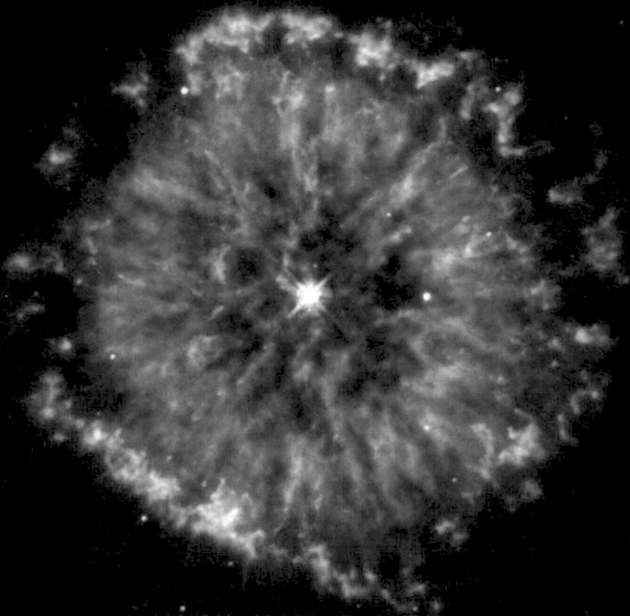

Astronomers using NASA's Hubble Space Telescope have obtained images of the strikingly unusual planetary nebula, NGC 6751. Glowing in the constellation Aquila like a giant eye, the nebula is a cloud of gas ejected several thousand years ago from the hot star visible in its center.

Hubble Heritage Project snapped this picture of NGC 1999, an example of a reflection nebula. Like fog around a street lamp, a reflection nebula shines only because the light from an imbedded source illuminates its dust; the nebula does not emit any visible light of its own.

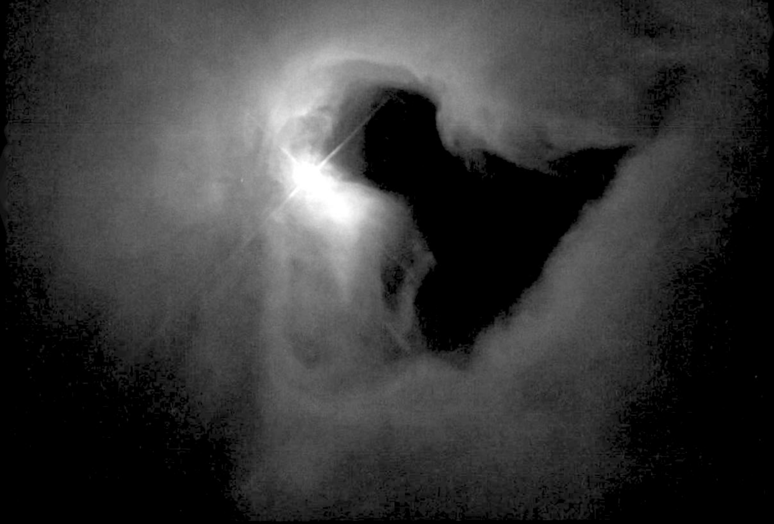

This Hubble image shows the unique galaxy pair called NGC 3314. Through an extraordinary chance alignment, a face-on spiral galaxy lies precisely in front of another larger spiral. This line-up provides the rare chance to visualize dark material within the front galaxy, seen only because it is silhouetted against the object behind it.

The larger and more massive galaxy (below) is cataloged as NGC 2207, and the smaller one (opposite page) is IC 2163. Strong tidal forces from NGC 2207 have distorted the shape of IC 2163, flinging out stars and gas into long streamers stretching out a hundred thousand light-years toward the right-hand edge of the image. (Continued on opposite page.)

In the year 1054 A.D., Chinese astronomers observed a new star, so bright that it was visible in broad daylight for several weeks. Today, the Crab Nebula (below) is visible at the site of the "Guest Star." Located about 6,500 light-years from Earth, the Crab Nebula is the remnant of a star that began its life with about 10 times the mass of our Sun.

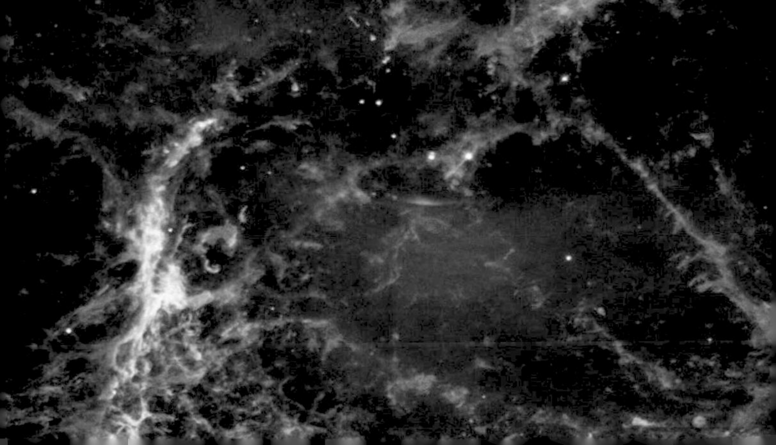

Hubble captured this image of the "butterfly wing"- shaped nebula, NGC 2346. The nebula is about 2,000 light-years away from Earth in the direction of the constellation Monoceros. It represents the spectacular "last gasp" of a binary star system at the nebula's center.

Hubble has allowed astronomers to resolve, for the first time, hot blue stars deep inside an elliptical galaxy. The swarm of nearly 8,000 blue stars resembles a blizzard of snowflakes near the core (lower right) of the neighboring galaxy M32, located 2.5 million light-years away in the constellation Andromeda.

This stellar swarm (below) is M80 (NGC 6093), one of the densest of the 147 known globular star clusters in the Milky Way galaxy. Located about 28,000 light-years from Earth, M80 contains hundreds of thousands of stars, all held together by their mutual gravitational attraction.

Penetrating 25,000 light-years of obscuring dust and myriad stars, Hubble has provided the clearest view yet of one of the largest young clusters of stars inside the Milky Way galaxy, located less than 100 light-years [illegible] the very center of the Galaxy. Having the equivalent mass greater than 10,000 stars like our sun, the monster cluster is ten times larger than typical young star clusters.

The majestic spiral galaxy NGC 4414 was imaged by the Hubble in 1995 using the Wide Field Planetary Camera 2 through three different color filters. Based on careful brightness measurements of variable stars in NGC 4414, astronomers were able to make an accurate determination of the distance to the galaxy of about 60 million light-years.

Hubble images of more than a dozen very distant colliding galaxies indicate at least in some cases, big massive galaxies form through collisions between smaller ones. This Hubble image shows paired galaxies very close together with streams of stars being pulled out of the galaxies.

This is a Hubble view of the spiral galaxy NGC 4603, the most distant galaxy in which a special class of pulsating stars called Cepheid variables have been found. It is associated with the Centaurus cluster, one of the most massive assemblages of galaxies in the nearby universe. The Local Group of galaxies, of which the Milky Way is a member, is moving in the direction of Centaurus at a speed of more than a million miles an hour.

This Hubble image is the extraordinary "polar-ring" galaxy NGC 4650A. Located about 130 million light-years away, NGC 4650A is one of only 100 known polar-ring galaxies. Their unusual disk-ring structure is not yet understood fully.

In this stunning picture of the giant galactic nebula NGC 3603, Hubble's crisp resolution captures various stages of the life cycle of stars in one single view. To the left of center is the evolved blue supergiant called Sher 25. The star has a unique circumstellar ring of glowing gas that is a galactic twin to the famous ring around the supernova 1987A.

A Hubble "family portrait" of young, ultra-bright stars nested in their embryonic cloud of glowing gases. The celestial maternity ward, called N81, is located 200,000 light years away in the Small Magellanic Cloud, a small irregular satellite galaxy of our Milky Way.

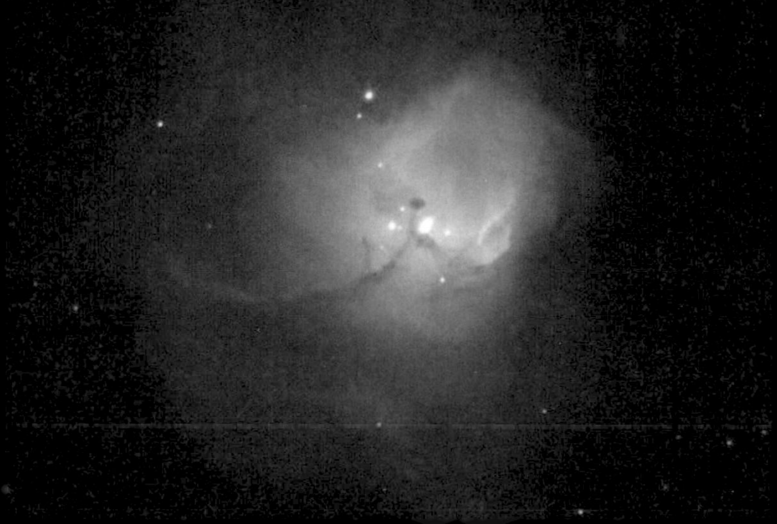

M2-9 is a striking example of a "butterfly" or a bipolar planetary nebula. Another more revealing name might be the "Twin Jet Nebula." If the nebula is sliced across the star, each side of it appears much like a pair of exhausts from jet engines.

In the most active starburst region in the local universe lies a cluster of brilliant, massive stars, known as Hodge 301. Seen in the lower right-hand corner of this image, it lives inside the Tarantula Nebula in the Large Magellanic Cloud. This star cluster is not the brightest, or youngest, or most populous star cluster in the Tarantula Nebula, that honor goes to the spectacular R136.

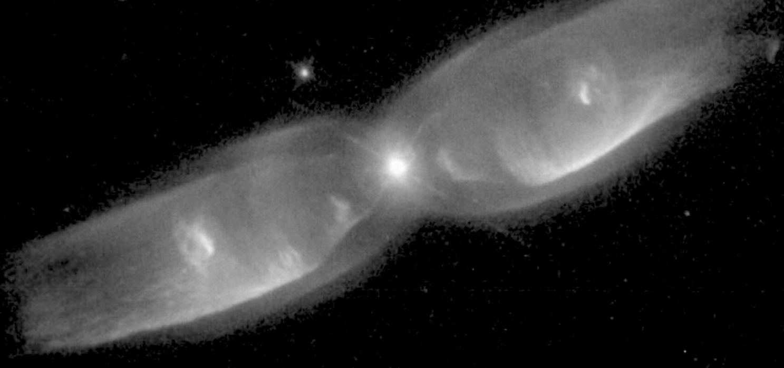

Glittering stars and wisps of gas create a breathtaking backdrop for the self-destruction of a massive star, called supernova 1987A, in the Large Magellanic Cloud. Astronomers in the southern hemisphere witnessed the brilliant explosion of this star on February 23, 1987.

Estimated to be 1,000 years old, one of the most complex planetary nebulae ever seen is NGC 6543 (below), nicknamed the "Cat's Eye Nebula." Hubble reveals surprisingly intricate structures including concentric gas shells, jets of high-speed gas and unusual shock-induced knots of gas.

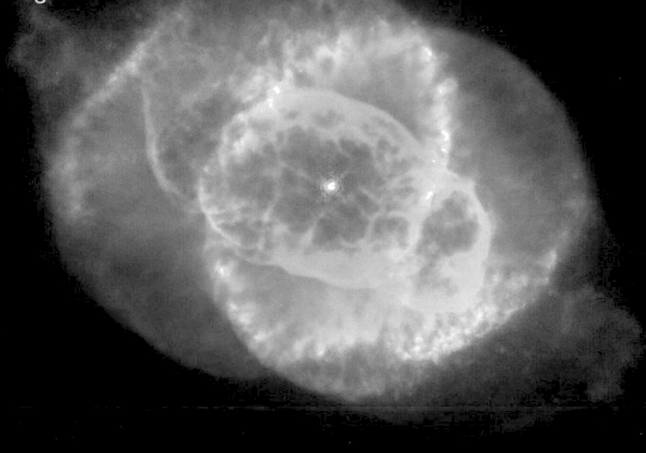

Hubble captured the sharpest view yet of the most famous of all planetary nebulae, the Ring Nebula (M57). In this image, the telescope has looked down a barrel of gas cast off by a dying star thousands of years ago. The nebula is about a light-year in diameter.

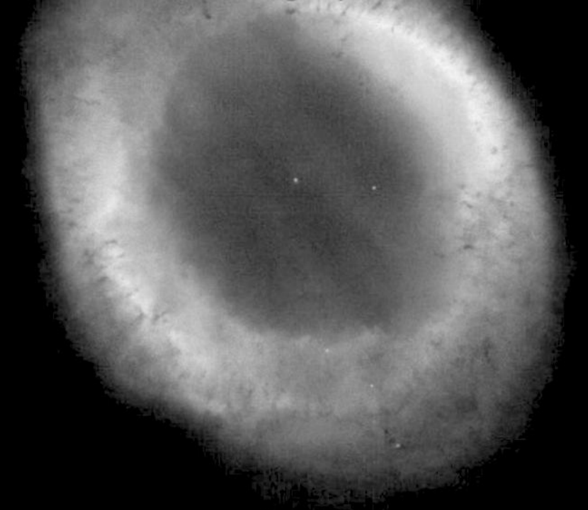

This Hubble image is of a vast nebula called NGC 604, which lies in the neighboring spiral galaxy M33, located 2.7 million light-years away in the constellation Triangulum. This is a site where new stars are being born in a spiral arm of the galaxy.

This image is a small portion of the Cygnus Loop supernova remnant, which marks the edge of a bubble-like, expanding blast wave from a colossal stellar explosion, occurring about 15,000 years ago. The image shows the structure behind the shock waves.

This comparison below of the core of the galaxy M100 shows the dramatic improvement in Hubble's view of the universe after the first Servicing Mission. The new image, taken with the second generation Wide Field and Planetary Camera installed beautifully demonstrates that the camera's corrective optics compensate fully for the optical aberration.

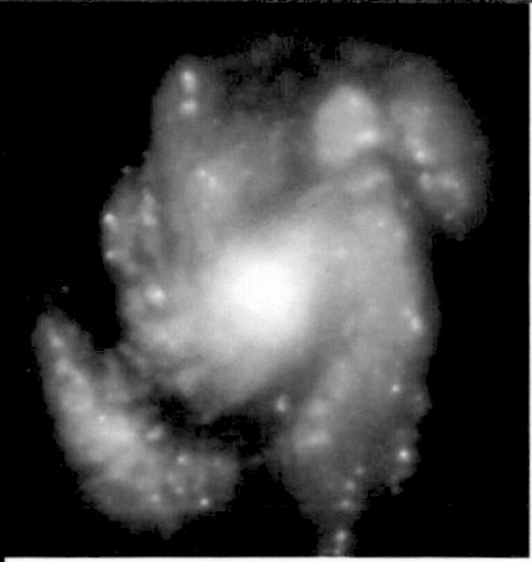

Wide Field Planetary Camera 1

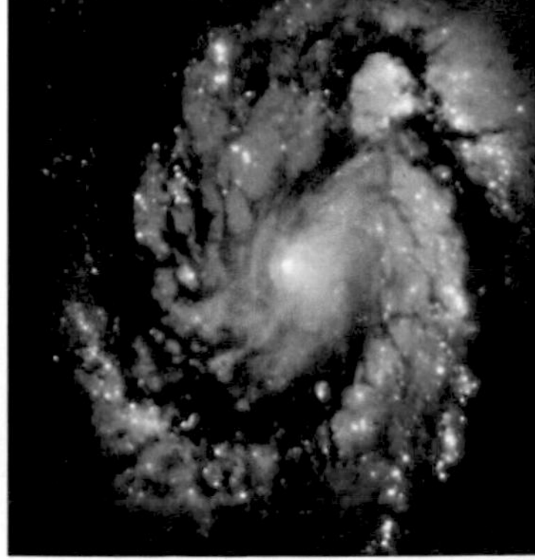

Wide Field Planetary Camera 2

This Hubble image reveals a pair of one-half light-year long interstellar "twisters," eerie funnels and twisted-rope structures in the heart of the Lagoon Nebula (Messier 8) which lies 5,000 light-years away in the direction of the constellation Sagittarius.

The flash from Star V838 Monocerotis "Echoes" Through Space. The red supergiant star in the center of this image brightened suddenly for several weeks in 2002, illuminating dust that may have been ejected from the star during a previous explosion.

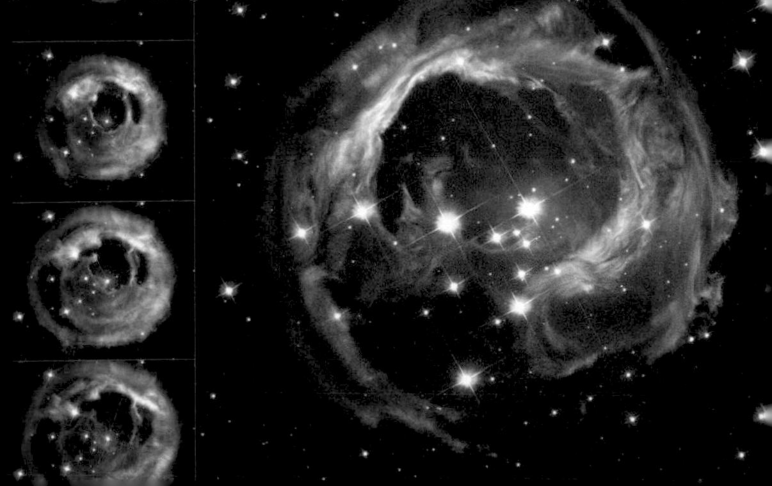

In spiral galaxy NGC 3370, intricate spiral arms contain hot areas of new star formation in this dusty galaxy. This galaxy was home to a supernova that appeared in 1994.

These columns of hydrogen gas and dust are star-forming pillars in the Eagle Nebula (M16) acting as incubators for new stars. The tallest pillar, on the left, is about four light years long.

The large Whirlpool Galaxy (M51) is known for its sharply defined spiral arms. Their prominence could be the result of the Whirlpool's gravitational tug-of-war with its smaller companion galaxy (right).

The Helix Nebula is a cloud of gas expelled and illuminated by the dying star at its center. From our vantage point on Earth it looks like a donut, but its actual structure is layered and complex.

The Sombrero Galaxy (M104) is a brilliant white core encircled by thick dust lanes in this spiral galaxy, seen edge-on. The galaxy is 50,000 light-years across and 28 million light years from Earth.

Clouds of dust and gas in the Orion Nebula churn out stars in this tiny section of the huge Orion Nebula. The gas is illuminated and heated by ultraviolet light from four hot, massive stars.

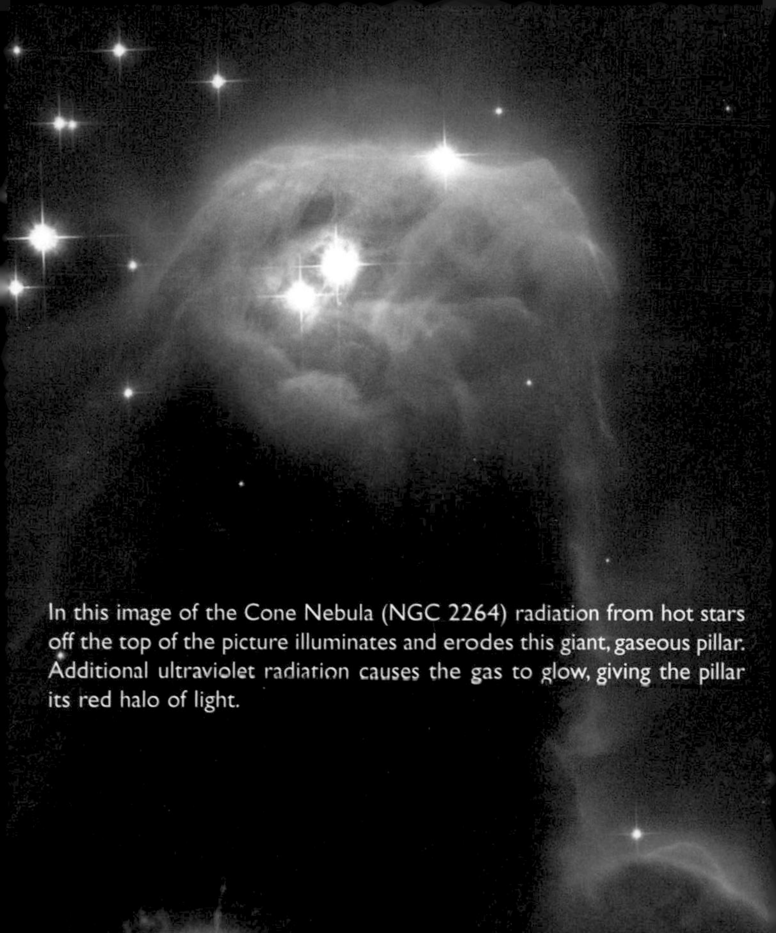

In this image of the Cone Nebula (NGC 2264) radiation from hot stars off the top of the picture illuminates and erodes this giant, gaseous pillar. Additional ultraviolet radiation causes the gas to glow, giving the pillar its red halo of light.

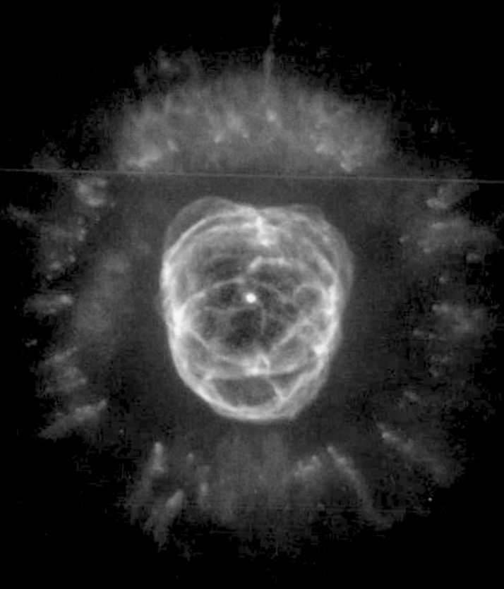

The Eskimo Nebula (NGC 2392) is a planetary nebula that began forming 10,000 years ago, when the dying star at its center began flinging out bubbles of gas and matter.

This is Hubble's sharpest view of the Orion Nebula. The Orion Nebula is a cavern of tumultuous gas and dust where massive stars are forming. The energy released by the young stars transforms their place of birth, whipping their surroundings into massive [illegible]

The Small Magellanic Cloud is a satellite galaxy of the Milky Way. It contains this group of baby stars that are still forming from collapsing gas clouds and have not yet ignited their hydrogen fuel.

First generation (inset) and second generation Tracking and Data Relay Satellites built by TRW and Boeing respectively.